T0075416

Contemporary Perspectives in Data Mining

Volume 4

Contemporary Perspectives in Data Mining

Series Editors

Kenneth D. Lawrence
New Jersey Institute of Technology

Ronald Klimberg
Saint Joseph's University

Contemporary Perspectives in Data Mining

Kenneth D. Lawrence and Ronald Klimberg, Series Editors

Contemporary Perspectives in Data Mining, Volume 4 (2021)
edited by Kenneth D. Lawrence and Ronald Klimberg

Contemporary Perspectives in Data Mining, Volume 3 (2017)
edited by Kenneth D. Lawrence and Ronald Klimberg

Contemporary Perspectives in Data Mining, Volume 2 (2015)
edited by Kenneth D. Lawrence and Ronald Klimberg

Contemporary Perspectives in Data Mining, Volume 1 (2015)
edited by Kenneth D. Lawrence and Ronald Klimberg

Contemporary Perspectives in Data Mining

Volume 4

edited by

Kenneth D. Lawrence
New Jersey Institute of Technology

and

Ronald Klimberg
Saint Joseph's University

INFORMATION AGE PUBLISHING, INC.
Charlotte, NC • www.infoagepub.com

Library of Congress Cataloging-in-Publication Data

CIP record for this book is available from the Library of Congress
http://www.loc.gov

ISBNs: 978-1-64802-143-5 (Paperback)

978-1-64802-144-2 (Hardcover)

978-1-64802-145-9 (ebook)

Printed in the United States of America

CONTENTS

SECTION III: BUSINESS APPLICATIONS OF DATA MINING

SECTION I

FORECASTING AND DATA MINING

CHAPTER 1

COMBINING FORECASTING METHODS

Predicting Quarterly Sales in 2019 for Motorola Solutions

Kenneth D. Lawrence and Stephan Kudyba
New Jersey Institute of Technology

Sheila M. Lawrence
Rutgers, The State University of New Jersey

COMBINING FORECAST METHODS WITH WEIGHTS

Often there is more than one forecast method of a given time series data source. The dilemma exists when considering which of these possible forecasting models should be chosen to produce the most accurate result. The simplest method for combining several forecasts is to average them but some assert that it may not be advisable to use simple averages when there is a large difference among the variances of the errors.

Contemporary Perspectives in Data Mining, Volume 4, pp. 3–9
Copyright © 2021 by Information Age Publishing
All rights of reproduction in any form reserved.

Research by Makridakis et al. (1982) however, has shown that simple averages produce more accurate forecasts than weighted averages when used with many time series. The concept of using weighting based on the inverse sum of squared errors was developed by Bates and Granger (1969).

If the forecaster has no preference for one forecasting method over another and if the variance of the errors do not differ dramatically, then the assumption of equal weight is a reasonable one. There are situations where difference weights could be based on other criteria.

The first of these method estimates is a weighted average that can be used to combine the forecasts. The higher the weights should be assigned to the forecast that has the lowest average.

The topic of the best ways to combine forecasts has been researched considerably. These studies have shown the following:

1.   In many cases, the process of combining forecasts of several models improve forecast accuracy as compared to a single forecast model.
2.   Simple unweighted models often perform **better** than more complex weighting structures.
3.   When one forecasting model is considerably **better** than others based on various statistical criteria, the advantage of combining forecasts is minimized. It is **better** to drop the inferior models from consideration.
4.   With weights that are determined by regression modeling analysis, the combination of the inverse of the sum of squares error is found to be inferior.

The combination of forecasts produces more accurate results than selecting an individual forecast (Clemen, 1989). A similar conclusion is found in Armstrong (2001). There is an element of insurance to using a number of forecasting methods. When market circumstances change, one method takes over from another as offering the best and most accurate forecasting. Ord (1988), provides a simple model that explains this effect.

When the accuracy measures of the forecasts are available, the higher weights should be given to the more accurate forecast models. The weights should be related inversely to the squared error of the forecast.

In general, the combination of a set of forecasts, based on a simple weighting scheme such as equal weights, produces more accurate forecasts than trying to select the best individual method. The fundamental reason that combining is so effective is that all methods suffer from differences. Even if a method is satisfactory for a period of time, the market or the economy will change, and then the method is likely to contain some information that it is helpful in improving the overall combined forecast. In addition, there is some element of insurance to using a number of forecasting method so that when market circumstances change, one method takes over from another as offering the best forecasts.

This is somewhat similar to adopting a portfolio of diverse assets in investment theory. Selecting investments with different market betas and risk parameters often helps reduce risk relative to returns given changes in economic and financial indicators and corresponding changes in asset prices.

For the most part, regression methods of combining forecasts are not as effective as weights determined unweighted sample averages or weights determined by the inverse of the sum of squares of errors.

Motorola Solutions Corporation

Motorola Solutions serves more than 100,000 public safety and commercial customers in more than 100 nations. As an industry leader, they design and develop communications devices and services to organizations in a variety of industries. The company focuses on developing integrated end-to-end solutions and products that empower seamless connectivity and their annual revenue has been in the range of 6 to 7.5 billion U.S. dollars over the past few years.

The company primarily works in the following areas:

1. Two-way radios
2. Command center software
3. Video surveillance and analytics
4. Public safety—long term evolution (LTE) 4GG technology devices and systems
5. Management and support services (intra structure; devices; cybersecurity)

The following analysis involves the forecasting of the quarterly sales of Motorola Solutions.

COMBINATION OF FORECASTING METHOD RESULTS

Seven different methods of time series forecasting were used to estimate one quarter ahead of the sales of Motorola systems, (e.g., first quarter 2019), based on eight years of quarterly sales data (2011–2018). The methods that were utilized are as follows:

1. Autoregressive modeling
2. Growth trend modeling

3. Linear trend modeling
4. Quadratic trend modeling
5. Single exponential smoothing
6. Double exponential smoothing
7. Moving average

The error in the forecast one quarter ahead of Motorola sales revenue is as follows (see Table 1.1):

Table 1.1

#	Forecasting Error	Modeling
1	−450.22	Autoregressive
2	280.31	Growth Trend
3	201.22	Linear Trend Modeling
4	−172.48	Quadratic Trend
5	−361.36	Single Exponential Smoothing
6	−486.46	Double Exponential Smoothing
7	−364.13	Moving Average

*errors were estimated as: (Actual rev (t + 1) − estimated (t + 1))

The MAPE (Mean Absolute Percent Error) of each of these models is as follows (Table 1.2):

Table 1.2

#	MAPE	Modeling
1	13.5	Autoregressive
2	**15.0**	Growth Trend
3	16.0	Linear Trend Modeling
4	13.8	Quadratic Trend
5	13.8	Single Exponential Smoothing
6	13.5	Double Exponential Smoothing
7	15.8	Moving Average

COMBINING OF FORECASTING MODELS OF THE NEXT QUARTER MOTOROLA SALES

Based on the magnitude of the error in forecasting, the following methods, based on the large error, will not be part of the combining forecast process:

- autoregressive modeling
- single exponential smoothing modeling
- double exponential smoothing modeling
- moving average modeling

Methods With Lower Errors Included in the Analysis

Linear Trend Forecasting Model

The model used is $Y_0 = b_0 + b_1 t$, where

Y_0 = is the Motorola sales revenue in period t
b_0 is the intercept of the linear trend line
b_1 is the slope of the linear trend line
t is the time period

Quadratic Trend Forecasting Model

The model used is $Y_t = b_0 + b_1 t, + b_2 t^2$ where
Y_t is Motorola sales revenue in period t
t is the time period

Growth Trend Forecasting Model

The model used in $Y_t = b_0 + (b_1)^t$ where
t is the time period

Forecasting Next Quarter Motorola Sales (First Quarter 2019), Based on Linear Trend, Quadratic Trending, Growth Trend

Method	Estimated Revenue
Linear Trend:	1,455.70
Quadratic Trend:	1,829.42
Quadratic Trend:	1,448.69

Combination of Forecasts, Based on Equal Weight of Motorola Sales First Quarter 2019

Assume Forecast (1Q) 2019 as Equal Weights $= 1,577.93$
Forecast Error $1,657.93 - 1,577.93 = 80.00$

Combination of Forecasting Based on MAPE Weighting of Motorola Sales Next Quarter (Table 1.3)

Table 1.3

Weighting	MAPE Weighting	Ratio
	$(MAPE_1)/$	
W_1:	$(MAPE_1 + MAPE_2 + MAPE_3)$	$16/436 = .38$
W_2:	$(MAPE_2)/$	
	$(MAPE_1 + MAPE_2 + MAPE_3)$	$13.8/43.6 = .31$
W_3:	$(MAPE_3)/$	$13.8/43.6 = .31$
	$(MAPE_1 + MAPE_2 + MAPE_3)$	

$.38\,(1,455.78) + .31\,(1,829.42) + .31\,(1,448.69)$

Forecast of (1Q) 2019:	Forecast Error (1Q) 2019:
$553.19 + 567.12 + 449.09 = 1,560.40$	$1,657 - 1,560.4 = 96.6$

Comparison of Results for 1Q2019

There is a significant difference in the magnitude of forecast error between the individual forecast error and both the combination of weighted methods of forecasting. The combination of forecasts for the equal weighted process is superior to the forecast error by the MAPE weighting process.

Implications of Work

Although the scenario tested in this work involves select data of one organization, the process of identifying superior modeling techniques is clearly depicted. When seeking to identify the most accurate modeling techniques, one method rarely fits all. Financial performance of organizations are unique to industry sector, company size and seasonality attributes, to name a few, that introduce a variety of variances that modeling methods must address.

This study examined just one approach to optimizing forecasting methods. There exist many other techniques and methods to consider.

REFERENCES

Armstsrong, J. S. (2001). Evaluating forecasting methods. In *A handbook for researchers and practitioners* (pp. 443–472). Klawer.

Bates, J. M., & Granger, C. (1969). The combination of forecasts. *Operational Research Quarterly, 20*, 451–468.

Clemen, R. T. (1989). Combining forecasts: A review and annotated Bibliography. *International Journal of Forecasting, 4*, 559–883.

Ord, J. K. (1985). Future developments in forecasting: The time series conversion. *International Journal of Forecasting, 4*, 389–402.

Makridakis, S., Andersen, A., Carbone, R., Fildes, R., Hibon, M., Lewandowski, R., Newton, J., Parzen, R., & Winkler, R. (1982). The accuracy of extrapolation. *Journal of Forecasting, 1*, 111–153.

CHAPTER 2

BAYESIAN DEEP GENERATIVE MACHINE LEARNING FOR REAL EXCHANGE RATE FORECASTING

Mark T. Leung
University of Texas at San Antonio

Shaotao Pan
Visa Incorporated

An-Sing Chen
National Chung Cheng University

ABSTRACT

With the strength of being able to apply probability to express all forms of uncertainty, Bayesian machine learning (ML) has been widely demonstrated with the capacity to compensate uncertainty and balance it with regularization in modeling. Simply put, this AI approach allows us to analyze data without explicit specification of interactions within parameters in a model. On the other hand, in the context of international economics, real exchange

Contemporary Perspectives in Data Mining, Volume 4, pp. 11–23
Copyright © 2021 by Information Age Publishing

rate has been used as a proxy measure of the relative cost of living and the well-being between two countries. A rise in one country's real exchange rate often suggests an escalation of national cost of living relative to that of another country. Thus, measuring and forecasting real exchange rates has profound implications to both economists and business practitioners. In this study, we use different forms of Bayesian machine learning to real exchange rate forecasting and comparatively evaluate the performances of these AI models with their traditional econometric counterpart (Bayesian vector autoregression). Empirical results indicate that, the tested Bayesian ML models (Boltzmann machine, restricted Boltmann machine, and deep belief network) perform generally better than the non-ML model. Although there is no clear absolute winner among the three forms of Bayesian ML models, deep belief network seems to demonstrate an edge over the others given our limited empirical investigation. In addition, given the rapid state-of-the-art advancement of this AI approach and its practicality, the current chapter also provides an abbreviated overview of the tested Bayesian machine learning models and their technical dynamics.

INTRODUCTION

With the strength of being able to apply probability to express all forms of uncertainty, Bayesian machine learning has been widely demonstrated with the capacity to compensate uncertainty and balance it with regularization in modeling. Simply put, this AI approach allows us to analyze data without explicit specification of interactions within parameters in a model. Instead of adopting a purely algorithmic process like traditional machine learning does, the process of Bayesian machine learning is carried out with respect to a probability distribution over all unknown quantities. When training is completed, a Bayesian machine learning model generates a complete posterior distribution, which yields probabilistic guarantees for predictions. At the same time, a prior distribution is used to integrate the parameters and to average across a spectrum of underlying models, achieving an effect of regularization of the overall Bayesian learning process. Essentially, Bayesian machine learning is a unique combination of machine learning and stochastic models with the stochastic portion forming the core of co-integration. It follows that Bayesian machine learning can produce probabilistic guarantees on its predictions and generate the distribution of parameters that it has learned from the observations. These two characteristics make this machine learning approach highly attractive to theoreticians as well as practitioners and lead to solid progress and adoption of the technique in the recent decade. Especially, popular machine learning methods (such as neural network, support vector machine, and relevance vector machine) have been extended to and encapsulated in the Bayesian paradigm. As a consequence, many Bayesian hierarchical structures have been developed. Applications of the developed Bayesian structures are numerous and some can be found in the areas of cancer screening, language modeling, and image caption generation. Interested readers can refer to Barber (2012) for more detailed exposition.

Due to the rich diversity within the environment of Bayesian machine learning, we limit our scope to three different forms of Bayesian deep generative models, namely, Boltzmann machine, restricted Boltzmann machine, and deep belief network. Although these learning machines are based on different architectural designs, they share a common statistical background and their logical mechanisms are interconnected, that is, one can be roughly viewed as a special form of the others. In light of these notions, the current chapter provides an overview of these models and their technical dynamics. The chapter also applies these three models to the forecasting of real exchange rates and compares their relative performances against a traditional yet promising econometric method—Bayesian vector autoregression.

In summary, the chapter is organized as follow. In the next section, we present an abbreviated overview of the three selected forms of Bayesian deep generative models for machine learning. Their working dynamics as well as their interrelationships are briefly explained. Subsequently, the macroeconomic framework and data (including preprocessing and transformation) used for real exchange rate forecasting are outlined. Results of empirical comparison of the three machine-learning and Bayesian vector autoregression (BVAR) models are reported and discussed in the following section. The final section concludes the chapter.

OVERVIEW OF BAYESIAN DEEP GENERATIVE MODELS

In the current study, we employ and analyze the performance of several modern architectures of Bayesian deep generative models (Kingma et al., 2014). All tested models derive their respective parameters from intrinsic probabilistic distributions. However, it should be noted that some of these models also allow explicit expression the probabilistic distribution and can be represented as graphs or factors. Since some of the tested models only allow drawing samples for derivation of the probabilistic distribution, there is no direct way to describe some of these models in the forms of graphs. Therefore, in order to achieve a fair comparison, we select the probabilistic method to obtain/optimize model parameters.

Boltzmann Machine

Boltzmann machine was first developed by Hinton (2002). The original machine represents a network of symmetrically connected units, which are connected to each other through bidirectional links. These units always take on stochastic binary states. It follows that this stochastic behavior is adopted as a probability function of the status from neighbor units and then applies weights on the connections. The binary characteristic of units can be interpreted as the acceptance or rejection of the hypothesis of the domain whereas the weights reflect the weaker constraint of the two competing hypotheses. When weights are positive and one hypothesis is to be selected, the machine will select the hypothesis with higher probability. As mentioned above, the connections between two units are symmetrical and bidirectional. As a result, the weights are also symmetrical and with the same magnitude in both directions.

Boltzmann machine can also be viewed as a probabilistic representation of Hopfield net. The global state of the network can be described by "energy," which is a number to be minimized. Under the proper assumptions and the right conditions, each unit works towards minimizing the global energy. In those cases where some constraints are imposed on the inputs, the network would adopt the changes and minimize the energy by updating the configuration to align with the inputs. Moreover, the energy is a metric to evaluate the extent of violation of the hypothesis by the network. In other words, by minimizing the energy, the network gradually and iteratively satisfies the global constraints.

For a d-dimensional random binary input $\mathbf{x}$. The join distribution is:

$$P(\mathbf{x}) = \frac{\exp(-E(\mathbf{x}))}{Z} \tag{1}$$

where $E(\mathbf{x})$ is the energy function and Z is the standardization function to make sure $P(\mathbf{x})$ is a valid probability. And the energy function of the Boltzmann machine is:

$$E(\mathbf{x}) = -\mathbf{x}^T W \mathbf{x} - \mathbf{b}^T \mathbf{x} \tag{2}$$

where W is the weights and $\mathbf{b}$ is the bias.

The Boltzmann machine is of great importance if the network contains latent variables. Similar to multilayer perceptron, these latent variables can be seen as neurons in hidden layers. With these latent variables, the Boltzmann machine is a generic model for estimating the joint probability with discrete variables (Le Roux & Bengio, 2008).

If we consider input units $\mathbf{x}$ is consist of observed units $\mathbf{v}$ and latent units $\mathbf{h}$, then equation 2 can be decomposed into:

$$E(\mathbf{v}, \mathbf{h}) = -\mathbf{v}^T W \mathbf{v} - \mathbf{v}^T U \mathbf{h} - \mathbf{h}^T V \mathbf{h} - \mathbf{b}^T \mathbf{v} - \mathbf{c}^T \mathbf{h} \tag{3}$$

To minimize the energy, the training method aims to find the local minimum of the weights and bias such that, when switching the on-off status of the units, it maintains the combinations for whichever unit that reduces the global energy. When applying maximum likelihood method to optimize the global joint probability, the updates on the connection between any two units is independent from other units. Therefore, the machine can be trained and the embedded weights can be optimized without knowing the status of other units, which implies that the learning rule is purely local. Other learning algorithms are also proposed. Most of these methods consider local status as well as other statistics. For example, Bengio (2015) combines back-propagation algorithm with energy-based machines and demonstrates the relationship to the continuous-latent-variable-based Boltzmann machine.

Restricted Boltzmann Machines

In light of the issues mentioned previously, restricted Boltzmann machine (Smolensky, 1986) also known as RBM was proposed. It is often considered as a bipartite graph with two layers—one layer consists of observed units which serves as the input layer while the other layer is made up of hidden layers for latent units. The visible layer and hidden layer are fully connected. However, within each layer, the units do not link to each other. Under this constraint, the unit in hidden layer is conditionally dependent on the visible layer. The two-layer architecture is the "signature" configuration for the RBM. This yields important building blocks for deep probabilistic models when more hidden layers are stacked in a systematic manner. More details regarding this aspect will be presented in the following section.

For a RBM with visible layer $\mathbf{v}$ and hidden layer $\mathbf{h}$, the joint probability distribution can be defined as:

$$P(\mathbf{v}, \mathbf{h}) = \frac{\exp\left(-E(v,h)\right)}{Z} \tag{4}$$

where the energy function is:

$$E(\mathbf{v}, \mathbf{h}) = -\mathbf{v}^T W \mathbf{h} - \mathbf{b}^T \mathbf{v} - \mathbf{c}^T \mathbf{h} \tag{5}$$

and Z is the partition function which serves a standardizing constant:

$$Z = \sum_v \sum_h \exp\left(-E(\mathbf{v}, \mathbf{h})\right) \tag{6}$$

Similar to the original Boltzmann machine, the RBM can be trained by updating all hidden units with the inputs from the visible units. Then update units in visible layer "reconstructively." These training steps are iterated until certain criteria are met.

One efficient algorithm proposed by Hinton (2002) is to use "contrastive divergence" quantity to approximate the gradient descent. First, the hidden layer is updated by observations from the real inputs. The updated hidden layer is then treated as a new "data" for training another RBM. The efficiency comes from treating one hidden layer at one time. By doing so, the algorithm also fits for the RBM with multiple hidden layers. After repeatedly following the procedure, the model is subsequently able to be fine-tuned later using back-propagation algorithm (Hinton & Salakhutdinov, 2006) or wake-sleep algorithm (Hinton et al., 2006)

Deep Belief Networks

Another type of stochastic generative models is deep belief networks (DBN), which usually consists of multiple latent hidden layers. Usually, these hidden layers are binary units. In a typical architecture of DBN, the connection between top two layers are not directional while the connections between the rest layers are directed. Like RBM, within each layer, the units are not connected. Every unit from any layers is only connected to its neighboring layers. If a DBN only contains two layers then it becomes an RBM.

The learning process of DBN usually involves two steps. The first step is to use Gibbs sampling to generate samples from the first two layers. In fact, we can consider this step as drawing samples from the RBM part in the DBN. The next step is to sample across the rest of the layers using ancestral sampling with one single pass (this is because these layers are not directly connected). At a particular instance, a DBN only deals with the units from one certain layer. Once the joint probability distribution is learned, it serves as the "data" for the subsequent layer. After this greedy layer-wise training, the model is still able to fine-tune all of the weights with wake-sleep algorithm (Hinton et al., 2006) to achieve a better overall generative performance.

The layer-by-layer training shows the property that units in one layer is dependent on the previous layer. In this top-down fashion, the generative weights are updated in an efficient manner. Conversely, if we think about the weights in the reverse direction, the bottom-up process helps to infer the value of the units in the latent layers.

However, there are still challenges related to DBNs. Because of the "explaining away" effect of the directed layer, it is sometimes difficult to interpret the posterior. Due to the bidirectional connection between the first two layers, the posterior depends on the prior and the likelihood. To get the prior of the first hidden layer, we have to integrate over the higher variables. In certain cases, evaluating the lower bound of the log likelihood is hard to achieve. Because the lower bound is proportional to the size of the width of the network, issues can be resulted from marginalizing the latent variables and defining the partition function for the top two layers.

Deep belief nets have been used for generating and recognizing images (Hinton, Osindero, & Teh, 2006; Ranzato et al., 2007; Bengio et al., 2007), video sequences (Sutskever & Hinton, 2007), and motion-capture data (Taylor et al., 2007). If the number of units in the highest layer is small, deep belief nets perform nonlinear dimensionality reduction and they can learn short binary codes that allow very fast retrieval of documents or images (Hinton & Salakhutdinov, 2006; Salakhutdinov and Hinton, 2007). In the next section, we will describe the data and the macroeconomic framework used for our exploratory experiment.

EXPLORATORY EMPIRICAL FORECASTING EXPERIMENT

Macroeconomic Framework

In order to examine the significance and extent of the models' predictive power, the three Bayesian ML models along with the traditional econometric BVAR model are applied to forecast the real exchange rates of Canadian dollar and British pound. For the sake of clarity, we should explicitly point out that the analytics in this study is based on *real* exchange rates rather than the *nominal* exchange rates widely assumed in majority of foreign exchange forecasting papers. In the context of international economics, real exchange rate (aka, real effective exchange rate) has been used as a proxy measure of the relative cost of living and the well-being between two countries. A rise in one country's real exchange rate often suggests an escalation of national cost of living relative to that of another country. Mathematically, the real exchange rate of a certain foreign currency can be computed by:

$$real\ exchange\ rate = (nominal\ exchange\ rate * CPI_{foriegn}) / CPI_{us} \quad (7)$$

where *CPI* is the consumer price index, which can be viewed as a proxy of the price level of the foreign or domestic country. Interested readers can refer to Evrensel (2013) for a quick review or Cushman (1983) for a detailed explanation from the international economic perspective.

To serve the purpose of forecasting, the macroeconomic framework of modified uncovered interest parity (MUIP) is adapted to the regularization paradigm of the Bayesian ML and BVAR models. Essentially, Sarantis and Stewart (1995a, 1995b) showed that the MUIP relationship can be modeled and written as follow:

$$e_t = \alpha_0 + \alpha_1(r_t^* - r_t) + \alpha_2(\pi_t^* - \pi_t) + \alpha_3(p_t^* - p_t) + \alpha_4(ca_t/ny_t) + \alpha_5(ca_t^*/ny_t^*) + \mu_t \quad (8)$$

where e is the natural logarithm of the *real* exchange rate, defined as the foreign (Canadian or British) currency price of domestic (U.S.) currency. r, π, p, and (ca/ny) are the natural logarithm of nominal short-term interest rate, expected price inflation rate, the natural logarithm of the price level, and the ratio of current account to nominal GDP for the domestic economy, respectively. Asterisks denote the corresponding foreign currency variables. μ is the error term. Sarantis and Stewart (1995a, 1995b) provide a more detailed discussion and investigation of this relationship.

Data

The time series data set used in our exploratory empirical forecasting experiment covers the period from January 1988 to December 2017. In order to perform proper in-sample model selection and out-of-sample forecasting, the complete set of observations are divided into three subsets: the training set (January 1988 to December 1997), the model selection/ determination set (January 1998 to December 2007), and the out-of-sample forecasting set (January 2008 to December 2017). The data split uses one-third of all data for model training and another one-third for model selection. Once the ML and BVAR models are appropriately specified and adjusted, they are retrained using two-thirds of entire data set (January 1988 to December 2007) to generate the out-of-sample forecasts. V-fold validation is also used to ensure the stability of the tested models and the intrinsic control parameters.

Original data are preprocessed and transformed to obtain the proper format necessary for MUIP-based forecasting. First, we transform the collected nominal exchange rates for U.S. dollar, Canadian dollar, and British pound into real exchange rate using equation (7). Second, we take the natural logarithm of all input and output variables in the data set. Third, we compute the lagged differences for each variable:

$$\Delta x_t = x_t - x_{t-1} \quad (9)$$

where x_t is an input or output variable observed at period t. Readers should be aware that the transformation specified in equation (9) is a modification to the original equation (8) in order to induce stationarity in the time series.

RESULTS AND DISCUSSION

The empirical experiment as described in the previous section are performed for the forecasting of the real exchange rates for Canadian dollars and British pounds. Performance statistics based on the out-of-sample forecasts from the four tested models are displayed in Tables 2.1 and 2.2. Specifically, conventional performance measures such as R-square (R^2), mean absolute error (MAE), root mean square error (RMSE), and U statistic are tabulated in Table 2.1 whereas the performance measures derived from Theil's Decomposition Test can be found in Table 2.2.

Table 2.1
Conventional Performance Statistics for Out-of-Sample Forecasts of Real Exchange Rates for Canadian Dollar and British Pound

Canada Dollar				
Model	R^2	MAE	RMSE	U Statistic
Bayesian Vector Autoregression (BVAR)	0.794	0.0341	0.0301	1.0000
Boltzmann Machine (BM)	0.874	0.0286	0.0269	0.8937
Restricted Boltzmann Machine (RBM)	0.898	**0.0242**	0.0253	0.8405
Deep Belief Network (DBN)	**0.903**	0.0258	**0.0215**	**0.7143**
British Pound				
Model	R^2	MAE	RMSE	U Statistic
Bayesian Vector Autoregression (BVAR)	0.724	0.0258	0.0275	1.0000
Boltzmann Machine (BM)	0.845	0.0172	0.0245	0.8909
Restricted Boltzmann Machine (RBM)	0.896	**0.0154**	0.0205	0.7455
Deep Belief Network (DBN)	**0.912**	0.0156	**0.0198**	**0.7200**

Table 2.2
Theil's Decomposition Test for Out-of-Sample Forecasts of Real
Exchange Rates for Canadian Dollar and British Pound

	Canadian Dollar			
Model	Bias Coefficient (a)	Significance Level of (a)	Regression Proportion Coefficient (b)	Significance Level of (b)
Bayesian Vector Autoregression (BVAR)	0.2046	0.1498	0.8535	0.0474*
Boltzmann Machine (BM)	0.0760	0.0428**	0.9469	0.0174*
Restricted Boltzmann Machine (RBM)	0.0879	0.0527	**0.9628**	0.0129*
Deep Belief Network (DBN)	**0.0637**	0.0350**	0.9593	0.0148*

	British Pound			
Model	Bias Coefficient (a)	Significance Level of (a)	Regression Proportion Coefficient (b)	Significance Level of (b)
Bayesian Vector Autoregression (BVAR)	0.1787	0.1061	0.8149	0.0972
Boltzmann Machine (BM)	0.0860	0.0473**	0.9265	0.0259*
Restricted Boltzmann Machine (RBM)	**0.0479**	0.0269**	0.9396	0.0197*
Deep Belief Network (DBN)	0.0537	0.0318**	**0.9623**	0.01542*

* indicates that the bias coefficient (a) is significant different from zero, that is, the null hypothesis of H_o: $a = 0$ is rejected at 5% significance level.

** indicates that the regression proportion coefficient (b) is significant different from one, that is, the null hypothesis of H_o: $b = 1$ is rejected at 5% significance level.

Based on all conventional performance statistics (see Table 2.1), the three tested Bayesian ML models outperform the traditional econometric BVAR model. It can be easily observed that (even without statistical tests) the R^2 of the ML models are significantly higher while their MAE and RMSE are significantly lower than the corresponding measures tied to BVAR forecasts. This observation is true for forecasting of both Canadian dollars and British pounds.

Since MAE and RMSE are not unit-free, these straightforward measures cannot be directly compared across dissimilar categories. Hence, we adopt U statistics (Cox & Hinkley 1974) to alleviate this deficiency. In short,

U statistic is defined as the ratio of the RMSE of a specified model to the RMSE of a benchmarking model. For our experiment, the numerator is the RMSE of each of the Bayesian ML models and the denominator is the RMSE of the benchmarking BAVR model. This arrangement represents a proxy of the improvement in forecasting power due to use of machine learning over the more traditional technique. As shown in Table 2.1, all U statistics for both exchange rates are significantly lower than 1, suggesting the superiority of the Bayesian ML models.

For the sake of clarity, the best performer in each measure is highlighted. The out-of-sample forecasting performances of the three forms of Bayesian ML models vary. In some cases, the performance gap can be negligible. Although there is no clear absolute winner among them, DBN seems to demonstrate an edge over the other two given the limited scope of our empirical investigation.

In addition, we also apply Theil's Decomposition Test to examine the relative forecasting strength of the ML models. Essentially, Theil's Decom-position Test is conducted by regressing the actual observed real exchanged rate, e_t, on a constant and the forecasted real exchange rate, $\hat{e}_t$, estimated by a particular model:

$$e_t = a + be_t \qquad (10)$$

The constant a (bias coefficient) should be insignificantly different from zero and the coefficient b for predicted real exchange rate (regression pro-portion coefficient) should be insignificantly different from the value of one (1) for the forecasting model to be acceptable (or deemed meaningful).

Table 2.2 shows the results of Theil's Decomposition Test on the out-of-sample forecasts generated by the four tested models. Similar to the conclusions drawn in Table 2.1, the Bayesian ML models yield statistically better bias coefficients (a) and regression proportion coefficients (b) for both Canadian and British real exchange rates. All constants and coeffi-cients estimated by Decomposition Test except in one case are significant at the 5% level. It should be pointed out that, although the bias coefficient (a) of RBM is not strictly statistically significant at 5%, its p-value of 0.0527 is very close to the discernable threshold. Following Table 2.1, the best performer in each measure is highlighted. The out-of-sample forecasting performances of the three forms of Bayesian ML models vary although the best values are from either RBM or DBF, depending on the exchange rate. Together with the observations from Table 2.1, one of the conclusions here, that is, the variance in relative performance, is to adopt at least one ML models for forecasting. Nevertheless, identification and development of remedies to resolve this issue is, for the moment, beyond the scope of this chapter.

CONCLUSIONS

In this chapter, we provide an abbreviated overview of Bayesian deep generative models in machine learning. Particularly, the exposition focuses on three selected forms—Boltzmann machine, restricted Boltzmann machine, and deep belief network. Their common background and inter-relationships are also explained. Subsequently, an exploratory empirical experiment is conducted in order to examine the significance and extent of the models' predictive power. The three ML models along with the traditional econometric BVAR model are applied to forecast the real exchange rates of Canadian dollar and British pound.

Empirical outcomes suggest that the three Bayesian ML models generate better out-of-sample forecasts than BVAR in terms of conventional forecast accuracy measures and Theil's Decomposition Test. Although the performances of the ML models vary, the better models strictly in our context are RBM and DBN. This calls for identification and examination of remedies to resolve or even take advantage of the issue. We should stress that our exploratory investigation is limited in scope (such as time horizon, number of exchange rates, inclusion of alternate training algorithms, etc.) and that the results reported in this chapter are considered more preliminary than definitive.

Future work can expand the time horizon and the number of exchange rates, in conjunction of including alternate training algorithms proposed in recent literature. Another angle of future research can focus on the use of bootstrapping to create more effective training paradigms as well as more comprehensive evaluation procedures. Finally, extended work can further include how to strategically combine ML modules in an array through the use of cloud technology.

REFERENCES

Barber, D. (2012). *Bayesian reasoning and machine learning*. Cambridge University Press.

Bengio, Y. (2015). *Early inference in energy-based models approximates back-propagation*. Technical Report. arXiv:1510.02777, Universite de Montreal.

Bengio, Y., Lamblin, P., Popovici, P., & Larochelle, H. (2007). Greedy layer-wise training of deep networks. *Advances in Neural Information Processing Systems 19*. MIT Press.

Cox, D. R., & Hinkley, D. V. (1974). *Theoretical statistics*. Chapman and Hall.

Cushman, D. O. (1983). The effects of real exchange rate risk on international trade. *Journal of International Economics, 15*, 45–63.

Evrensel, A. (2013). *International Finance for Dummies*. John Wiley. https://www. dummies.com/education/finance/international-finance/what-are-real-exchange-rates/

Hinton, G. E. (2002). Training products of experts by minimizing contrastive divergence. *Neural Computation, 14*, 1711–1800.

Hinton, G. E, Osindero, S., & Teh, Y. W. (2006). A fast learning algorithm for deep belief nets. *Neural Computation, 18*, 1527–1554.

Hinton, G. E., & Salakhutdinov, R. R. (2006). Reducing the dimensionality of data with neural networks. *Science, 313*, 504–507.

Kingma, D. P., Mohamed, S., Rezende, D. J., & Welling, M. (2014). Semi-supervised learning with deep generative models. In *Proceedings of/Advances in Neural Information Processing Systems* 27 (NIPS 2014), Montreal, PQ.

Le Roux, N., & Bengio, Y. (2008). Representational power of restricted Boltzmann machines and deep beliefs network. *Neural Computation, 20*, 1631–1649.

Ranzato, M., Huang, F. J., Boureau, Y., & LeCun, Y. (2007). Unsupervised learning of invariant feature hierarchies with applications to object recognition. In *Proceedings of Computer Vision and Pattern Recognition Conference (CVPR 2007), Minneapolis, MN.*

Salakhutdinov, R. R. and Hinton, G. E. (2007). Semantic Hashing. In *Proceedings of the SIGIR Workshop on Information Retrieval and Applications of Graphical Models, Amsterdam, Netherland.*

Sarantis, N., & Stewart, C. (1995a). Monetary and asset market models for sterling exchange rates: A cointegration approach. *Journal of Economic Integration, 10*, 335–371.

Sarantis, N., & Stewart, C. (1995b). Structural, VAR and BVAR models of exchange rate determination: A comparison of their forecasting performance. *Journal of Forecasting, 14*, 201–215.

Sutskever, I., & Hinton, G. E. (2007). Learning multilevel distributed representations for high-dimensional sequences. In *Proceedings of AI and Statistics 2007, Puerto Rico.*

Taylor, G. W., Hinton, G. E., & Roweis, S. (2007). Modeling human motion using binary latent variables. *Advances in Neural Information Processing Systems* 19. MIT Press.

Welling, M., Rosen-Zvi, M., & Hinton, G. E. (2005). Exponential family harmoniums with an application to information retrieval. *Advances in Neural Information Processing Systems 17*. MIT Press.

CHAPTER 3

PREDICTING HOSPITAL ADMISSION AND SURGERY BASED ON FRACTURE SEVERITY

Aishwarya Mohanakrishnan,
Dinesh R. Pai, and Girish H. Subramanian
Penn State Harrisburg

ABSTRACT

According to World Health Organization, falls are the second leading cause of accidental injury deaths worldwide. In the United States alone, the medical costs and compensation for fall-related injuries are $70 billion annually (National Safety Council). Adjusted for inflation, the direct medical costs for all fall injuries are $31 billion annually of which hospital costs account for two-thirds of the total. The objective of this paper is to predict fall-related injuries that result in fractures that ultimately end up in hospital admission. In this study, we apply and compare Decision Tree, Gradient Boosted Tree (GBT), Xtreme Gradient Boosted Tree (XG Boost) and Neural Networks modeling methods to predict whether fall related

Contemporary Perspectives in Data Mining, Volume 4, pp. 25–38
Copyright © 2021 by Information Age Publishing
25

injuries and fractures result in hospitalization. Neural networks had the best prediction followed by XG Boost and GBT methods. By being able to predict the injuries that need hospital admission, hospitals will be able to allocate resources more efficiently.

INTRODUCTION

According to World Health Organization (WHO, 2018), falls are the second leading cause of accidental injury leading to deaths worldwide. In the United States alone, the medical costs and compensation for fall related injuries are $70 billion annually (National Safety Council, 2003). Adjusted for inflation, the direct medical costs for all fall related injuries are $31 billion annually, in which the hospital costs alone account for two-thirds of this total (Burns et al., 2016). According to the Centers for Disease Control and Prevention (CDC) and the National Center for Injury Prevention and Control's (NCIPC) Web-Based Injury Statistics Query and Reporting System (WISQARS, 2016), it can be seen that over 800,000 patients annually are hospitalized for fall-related injuries—mainly head injuries or hip fractures. Additionally, one out of five falls causes serious injuries, such as severe head injuries or broken bones (Alexander et al., 1992; Sterling et al., 2001). Owing to the fact that one-third of community-dwelling adults fall each year, preventive measures can be formulated by studying the trends of fall-related injuries to reduce the cost to the health systems in terms of morbidity and mortality (Shankar et al., 2017). It would also help hospitals to better allocate their resources such as operating rooms, beds, hospital rooms, doctors and nurses in accordance to the severity and duration of stay of the patients. While most research done so far has been limited to injuries related to older people in their homes, rehabilitation centers, or communities, our study will predict whether a fall-related injury in an emergency department setting will lead to admission or discharge of the patient.

REVIEW OF THE CURRENT KNOWLEDGE ABOUT HOSPITALIZATION FOR FALL AND FALL-RELATED INJURIES AND ITS PREDICTION

Falls among older people in institutionalized environments have become a growing concern. It has also been an interest of many research studies to reduce the substantial human and economic costs involved. National Electronic Injury Surveillance System All Injury Program (NEISS-AIP) results predicted that the annual cost of unintended fall-related injuries that lead to fatality, hospitalization, or treatment in an emergency department was $111 billion in 2010 (Verma et al., 2016). Diagnosis at the time of preadmission assessment predicted the legal status, the triage security and urgency scale, which ideally helped in predicting a patient's length of stay, as beds in hospitals are expensive (Orces & Alamgir, 2014). Patient's age, sex, duration of presenting problem, diagnosis, laboratory and radiographic investigations, treatments, and referrals were the main indicators in the triage to categorize patients into "Primary Care" and "Accident and Emergency" in a study conducted with 5,658 patients in the accident and emergency department in a hospital in London (Dale et al., 1995). Proximal Humerus Fracture was predicted using 2,85,661 patient samples, in which 19% of the cases were admitted and 81% of the cases were discharged from the Nationwide Emergency Departments between the years 2010 and 2011 (Menendez & Ring, 2015). The authors identified that factors such as Charlson Comorbidity Index, injury due to motor vehicle crash, polytrauma, ED visit on a weekday, Medicare and Medicaid insurance, open fracture, urban teaching hospital, increasing age and residence in the northeast were all associated with inpatient admission for Proximal Humerus Fracture (Menendez & Ring, 2015). Results from the National Health Interview Survey (NHIS) as well as the CDC and Prevention's Web-based Injury Statistics Query and Reporting System (WISQARS) showed that, of all the fall-related injuries in adults, 32.3% occurred among older adults (65+ years), 35.3% among middle-aged adults (45–64 years) and 32.3% among younger adults (18–44 years) (Verma et al., 2016.). The diagnostic category of a patient having more than one disease or symptom was found to be a statistically significant predictor of injurious falls (Amy et al., 2016). Patients who were screened for a fall-risk had a lower rate of fracture after a fall than those who were not screened (Chari et al., 2013). Further, it was also found that fractures were more likely to occur while the patients were walking or standing than during other activities in both settings. Another important predictor was the time periods from 14:00 and 15:00 and 21:00 and 22:00 hours in the hospital setting and 7:00 and 8:00, 16:00 and between 17:00 and 19:00 and 20:00 hours, which led to higher fracture odds (Chari et al., 2013).

Women had higher risk of falls than men (Dhargave & Sendhilkumar (2016). Other factors such as the history of falls, poor vision, use of multiple medications, chronic diseases, use of walking aids, vertigo, and balance problems led to higher fall rate among the elderly population living in long term care homes (Dhargave & Sendhilkumar, 2016). Similar to Chari et al.'s (2013) study, a higher rate of falls was found to occur during the day time (i.e., 72.5%—morning, 17.5%—afternoon, 5%—evening, 5%—night) (Dhargave & Sendhilkumar, 2016). Two most common injuries which led to fatalities were hip-fractures (54%) and head-injuries (21%) (Deprey et al., 2017).

The hospitalization rate and fracture-related injuries for elderly women was 1.8 times (i.e., 70.5% were women) and 2.2 times more than that of elderly men. Fractures, contusions/abrasions, and lacerations accounted for more than three quarters of the injuries in which rate ratios were greatest for injuries of arm/hand, leg/foot and lower trunk (Stevens & Sogolow, 2005). A simple linear regression model found that the rate of visits to the emergency among older adults aged 65 or more, increased from 60.4% in 2003 to 68.8% in 2010 per 1,000 population. And among the subgroups of patients, visits by patients aged 75–84 years increased from 56.2% to 82.1% (p <.01), visits by women increased from 67.4% to 81.3% (p = 0.04), visits by non-Hispanic Whites increased from 63.1% to 73.4% (p < 0.01), and visits in the South increased from 54.4% to 71.1% (p = 0.03) (Shankar et al., 2017).

Five clinically significant variables—history of falls in the last 12 months, history of loss of balance in the last 12 months, slowing of walking/change in gait, weak hand grip, and impaired sight were identified in (Miedany et al., 2011). The fall risk assessment score (FRAS) questionnaire offers a first line of risk assessment for all health care professionals to identify the patients who need further assessment, management and then assistance. FRAS can be used for inpatients and outpatients in an acute hospital setting thereby reducing the negative effects of falls on the elderly patients' social, physical and psychological functional abilities (Miedany et al., 2011). The statistically significant independent variables which helped identify/predict elderly osteoporotic women who had an increased fracture-risk were prior fracture, prior falls, lower BMD, and/or selected factors associated with falls (physical/cognitive dysfunction) (Weycker et al., 2017).

It is clear from these studies that fall related injuries happen and such patients most likely will be taken to emergency departments and hospitals. An effective prediction is needed to determine if hospitalization is needed in such cases to help in effective resource allocation at hospitals.

RESEARCH STUDY AND METHODOLOGY

Numerous studies in the past have successfully used classification algorithms such as the classification and regression trees in disease diagnosis and/or prognostic prediction varying from 5-year hip fracture risk (Jin et al., 2004) to short term risk of fractures among osteopenic women (Miller et al., 2004). A retrospective cohort study, conducted on a total of 29,848 Long Term Care (LTC) residents, where 22,386 cases consisted of the derivation dataset and 7,462 cases were taken as the validation dataset. The Fracture Risk Scale using decision tree analysis was developed to predict hip fracture over a 1-year time period, under eight different risk levels ranging from 0.6% (lowest risk level) to 12.6% (highest risk level) (Ioannidis et al., 2017). Decision Tree was particularly effective as it is a non-parametric algorithm and can efficiently deal with large, complicated datasets without imposing a complicated parametric structure (Song & Lu, 2015).

The clinical and radiological risk factors affecting pathological fractures was examined using logistic regression in 315 lung cancer patients with metastasis to the femur, between January 2010 and December 2014 (Eunsun et al., 2017). In this study, comparison analysis was done using five different machine learning algorithms namely, the AdaBoost, Support Vector Machine (SVM), Linear Discriminant Analysis (LDA), Random Forest and Gradient Boosting model (GBM), it was demonstrated that GBM was the best classifier for predicting the pathological femoral fractures. Another study of postmenopausal women, found that GBM when incorporated with diverse measurements of bone density and geometry from central QCT imaging and of bone microstructure from high-resolution peripheral QCT imaging, can improve fracture prediction (Atkinson et al., 2012). This study found that with all possible variables in the GBM model, the AUC was close to 1.0, predicting fracture and nonfracture with surprisingly high predictability. This study proves that more complex modeling techniques, such as, GBM creates stronger fracture predictions though there is a possibility that these models maybe overfitted, thereby producing results better than it would with new data added to the model (Atkinson et al., 2012).

Xtreme Gradient Boosting or Regularized Gradient Boosting model is a powerful implementation of the gradient boosting, first proposed by Friedman and Jerome (2001). This model was designed for speed and performance (Povalej Brzan et al., 2017). It is a machine learning system used for creating boosted trees, which gives us more accurate results than most machine learning techniques, as it avoids the overfitting problems by regularizing the data (Chen & Guestrin, 2016). XG Boost is a success in varied predictions problems, such as store sales prediction, high energy physics event classification, web text classification, customer behavior prediction, motion detection, ad click through-rate prediction, malware classification, product categorization, hazard risk prediction, and massive online course dropout rate prediction. (Chen & Guestrin, 2016). XGBoost was used in a study to predict hip fractures and estimate the predictor importance in dual energy X-ray absorptiometry scanned men and women in two Danish regions between 1996 and 2006 combined with Danish patient data which comprised of 4,722 women and 717 men with a 5-year follow-up time (Kruse et al., 2017). Among the 24 statistical models in this study, eXtreme Gradient Boosting performed the best to predict the fracture-rate in men. The findings of this study proved that machine learning techniques can improve the prediction beyond logistic regression, using ensemble models (Kruse et al., 2017). In another study, the California Statewide inpatient database was used to build regularized logistic regression models to predict readmission of morbidly obese patients ($n = 18,881$). XG Boosting used to predict and validate the prediction gave results with high accuracy as expected (Povalej Brzan et al., 2017).

Artificial Neural Networks (ANNs) excel in pattern recognition using previously solved examples to build a system of neurons to make new decisions, classify and forecast (Greenwood, 1991: Terrin et al., 2003). One such study which consisted of 286 cases of hip-fracture surgery, from the Department of Orthopedics, National Taiwan University Hospital Yun-Lin Branch, predicted 20–30% mortality rate of older patients one-year after the surgery. It was found that ANNs have a higher predictive ability than logistic regression mainly because they are not affected by the interaction between factors, which is why they will be hugely helpful in assisting complex decision making in a clinical setting (Lin et al., 2010).

Another prospective study of 18,362 patients undergoing cardiac surgery used ANNs to select a minimal set of risk variables to predict mortality with high accuracy. The neural networks selected 34 relevant risk factors to predict mortality and the results obtained were superior to those produced by the logistic regression models (Nilssonet al., 2006). A similar study was done on 1,167 women aged 60 years and above with an incidence of new hip-fracture during a period of 10 years. The findings indicated that ANNs were able to predict hip fractures more accurately than any of the existing statistical models. It can help stratify individuals in a clinical management (Ho-Le et al., 2017).

In this study, we apply and compare Decision Tree, GBT, Xtreme Gradient Boosted Tree (XGBoost) and neural networks modeling methods to predict whether fall related injuries and fractures result in hospitalization.

DATA AND VARIABLES

Data

The retrospective patient fracture data pertains to the emergency department of a nonprofit hospital group in Central Pennsylvania. The emergency department (ED) in the hospital is not an accredited trauma center and they function 24 x 7 with the resources optimized according to the in-flow of patients. The data set consists of patient information collected between January 1, 2013 and September 30, 2016. The original dataset contained 29,174 rows, that is, patient visits to the emergency department. Data pertaining to patients 18 years and younger were discarded as our study focuses only on adult patients. The final dataset contained 23,924 observations.

Variables

We define below the variables included in our analysis in this study. The hospital's emergency department has following data collection process. For a patient presenting to the ED, among others, the hospital's information system collected following variables: *Triage date*: The date the patient was registered as an emergency department patient; *Greet time*: An ED personnel greeted patients as they entered the ED and recorded the time of arrival. *Patient age:* Age of the patient was noted along with proper units such as days, months, and years. We consider only adults in this study. Furthermore, we converted the variable age into categorical variables: 19–44-years, 45–64 years, 65–74 years, 74–85 years, and 85 plus-years to derive more insights on the impact of age on falls. Each of these variables were treated as binary variables. *Patient gender*: We created a dummy variable "Gender" (female = 1, and male = 0); *Patient race*: Patient race was considered by the treating provider when the illness of injury may be prevalent to a specific race. We created three dummy variables—African American, non-Hispanic Whites, and Others, to determine the impact of race on falls; *Arrival mode*: The patients presenting to the ED typically arrived from home or by an ambulance. Approximately 55% of the patients arrived by an ambulance. We created a dummy variable "Arrival Mode" (ambulance = 1; home = 0); *Chief complaint*: This is the injury voiced by the patient; however, if a specific injury or illness is not voiced, the emergency department staff completed this field with a description that most accurately noted the injury or illness; *Emergency severity index* (ESI): This is a tool used during the triage process in emergency departments to assess acuity and resources that will be needed to treat the patient. It ranges from ESI 1 to ESI 4, where ESI 1 indicates the most urgent patient and ESI 5 indicates the least resource intensive. We created five binary variables to determine the impact of severity on falls; *Diagnosis*: This indicates the primary diagnosis of the patient upon discharge, admission or transfer. Patients may have multiple conditions that require clinical intervention during the episode of care; however, the diagnosis notes the overriding condition. This may differ from the chief complaint. For example, a patient may have been involved in a car accident, chief complaint would be listed as MVA (motor vehicle accident). The diagnosis may be listed as myocardial infarction. In this case, the patient had a cardiac event that caused the driver to lose control and have a car accident; *Disposition Code*: In this study we considered two types of disposition: discharge (DC)—patient was discharged from the emergency department upon completion of treatment, and hospitalization, i.e. inpatient (INPT)—patient was admitted to the hospital on an inpatient unit. Disposition code (0,1) is the dependent variable in the model (discharge = 1 and 0 = inpatient or hospitalization).

The dependent variable for the model was taken as the "Disposition Code" and the independent variables are shown in Table 3.1.

Table 3.1
Independent Variables for the Models Used in this Research

Independent Variables					
Abrasion	Brain	Femur	Heart	Others	Spine
AgeGroup	Cellulitis	Ffall	Hip	Pelvis	Stroke
Alcohol	Chest	Fibula	Knee	Pneumonia	Syncope
AmbDys	Concussion	FirstUrgency	Laceration	TriageDay	Tibia
Anemia	Contusion	Foot	Leg	TriageMonth	TimeToGreetHour
Ankle	Dehydration	ForeArm	Lumbar	Renal	UpperArm
AnklePain	Diabetes	Fracture	Mental	Rhabdomyolysis	UTI
ArrivalBy	Dizzy	Gender	Nasal	Rib	Weakness
Back	Elbow	Hand	Neck	Scalp	Wrist
Blood	Fall	Head	OffHours	Shoulder	

Results

The data was split into "Training" dataset and "Testing" dataset in the ratio of 80% and 20% respectively. Though it was a random split, the percentage distribution of the dependent variable in both "Training" and "Testing" datasets were similar (67% discharges and 33% inpatients). The Decision Tree model was trained based on 80% of the data and tested on the remaining 20%. The model accuracy should be similar in both the "Training" and "Testing" datasets, to make sure there is no overfitting of the model. The splitting criteria is taken as the "Gini Index," since majority of the variables in the dataset have only 2 classes (Binary Splits). GBM is constructed with the model's parameters, such as the number of trees which is taken as 2,000 and learning rate which is taken as 0.01. The train fraction is taken as 0.75, to train on 75% of the data and validate on the rest 25% of the data. At around 2,000 trees, both training and validation deviance is the least. Hence, the best tree chosen by the model is 1999. XG Boosting is a boosting based model, which works very well for classification problems as it trains the dataset by sequential learning thereby reducing the misclassification rate tree by tree. While this is similar to the GBM, the main advantage of using XG Boost is that it uses regularization, which penalizes variables that are not contributing to the model. So, it helps reduce the overfitting problem prevalent in many of the machine learning techniques. XG Boost does not work on missing values (NA). Since our data does not have any NAs it was not a problem. The model only works on numeric columns hence, the categorical fields were converted into numeric values using one hot encoding. It works on matrix data formats and not data-frames therefore, the data was converted into a matrix format. Our objective is chosen as binary logistic since we are dealing with a classification problem. The ETA is step size shrinkage used in update to prevent overfitting. After each boosting step, we can directly get the weights of the new features and the ETA actually shrinks the feature weights to make the boosting process more conservative. We gave 0.3 to reduce overfitting and the max-depth was given as 6 to capture any nonlinearity. The boosting was chosen to be run for 100 rounds with a 5-fold cross validation. Neural networks help us identify all kinds of relationship between dependent and independent variables. It can give us the highest possible accuracy and it works well for cases with high number of features. 100 hidden layers were chosen corresponding to the high number of input variables (108 input variables) and linear output was turned off since it is a binary classification problem. The model is then run and the following ROC curves are seen.

To compare the results from these models, we need certain evaluation metrics or measures.

Sensitivity: It is also known as True Positive Rate, Recall or the probability of detection. This gives us a measure of the proportion of the positives identified correctly, that is, "When, it's actually a yes, how often does the model predicts a yes?" Among all the models we have run or implemented in this study, Neural Networks has the highest sensitivity of 0.99 for the "Training" dataset and 0.98 for the "Testing" dataset.

Specificity: It is the measure of the proportion of the negatives that are correctly identified, that is, "When, it's actually a no, how often does the model predicts a no?" It is the equivalent to True Negative Rate. Looking at Table 3.2, we can see that, Neural Networks has the highest specificity of 0.91 for "Training" dataset and 0.86 for the "Testing" dataset.

Precision: It is also known as the Positive Predictive Value. It is basically, the fraction of relevant instances among the retrieved instances, that is, "When, the model predicts a yes, how often is it correct?" The Tables 3.2 and 3.3 show that, among all the models used in this study, Neural Networks has the highest precision of 0.96 for the "Training" dataset and 0.94 for the "Testing" dataset.

F1 score: In statistical analysis as the ones done in this chapter, the F1 score (also known as, the F1 measure), is a measure of a test's accuracy. It is the harmonic mean of the True Positive Rate and Precision. It helps us measure the performance of the models, in terms of both recall and the precision, where an F1 score reaches its best at the value 1 and worst at value 0. Hence, it can be seen from the above tables, that Neural Networks has the highest F1 score of 0.98 for the "Training" dataset and 0.96 for the "Testing" dataset.

Table 3.2
Evaluation Metrics of Training Dataset for All the Models

Evaluation Parameters	Decision Tree	GBM	XG Boost	Neural Networks
Precision	0.858	0.902	0.936	0.961
Specificity	0.688	0.789	0.865	0.917
Sensitivity	0.914	0.940	0.966	0.999
F1 score	0.885	0.920	0.951	0.980

Table 3.3
Evaluation Metrics of Testing Dataset for All the Models

Evaluation Parameters	Decision Tree	GBM	XG Boost	Neural Networks
Precision	0.853	0.904	0.924	0.935
Specificity	0.686	0.802	0.845	0.864
Sensitivity	0.913	0.936	0.953	0.989
F1 score	0.882	0.920	0.938	0.962

The evaluation metrics seen so far, shows us that Neural Networks give us superior prediction rates with high accuracy. It can also be seen from the above tables, that they have a consistent performance, on "Training" as well as "Testing" datasets, which gives us the confidence to reuse the model on any external dataset. Decision Tree had the lowest scores on all these measures, since it is a simple learning model based on just one tree. Hence, it is evident that adding many learning layers (like GBM, XG Boost, and Neural Networks) help us improve the accuracy drastically.

DISCUSSION

Our chapter deals with a classification problem of whether a patient will be admitted or discharged after treatment in an ED. Triangulation of results obtained from using multiple prediction models is vital for predicting and understanding fractures that result in hospital admissions. Hence, we first start by using tree-based algorithms such as the Decision Tree, the Gradient Boosted Tree, and the Xtreme Gradient Boosted Tree.

Decision Trees are simple and intuitive, yet powerful algorithms, which helps us identify the reasons or drivers behind the classification. After the initial analysis using the decision tree proved fruitful, the next step was to use a boosting algorithm to improve the accuracy of the model.

GBM was chosen as the boosting algorithm, to increase the framework's efficiency. GBM is nothing but a forest of many sequential decision trees. It is important to tweak parameters such as number of trees, validation ratio, sample size, minimum observations in a node, and so forth. Optimizing the performance by trying different combinations of parameters in a trial and error method was time consuming and performance intensive task. GBM proved to be much better than the Decision Tree model.

To further improve the results, we used another boosting algorithm. The main advantage of the Xtreme Gradient Boosting model is that it is not prone to overfitting. But the problem with this model is that it is computationally intensive and time consuming. Another drawback is that it takes only "numeric" variables. Hence, all factor variables had to be changed to numeric ones which create a really wide table. The XG Boost final model gave much better accuracy than the GBM.

Next, the forward pass of a Neural Network was done using 100 hidden layers with the sigmoid function to learn as much as possible from the data. The prediction rate of the final model was much better than the tree-based models. The neural network model was performance intensive and time consuming. While the tree-based algorithms give us intuitive variable importance measures, the variable importance generated by the Neural Networks are difficult to comprehend. Hence, it can be used where prediction accuracy is more important than the interpretability of the model.

CONCLUSION AND FUTURE RESEARCH

The objective of this paper is to predict fall-related injuries that result in fractures that ultimately ends up in hospital admission. We compare Decision Tree, Gradient Boosted Tree (GBT), Xtreme Gradient Boosted Tree (XG Boost), and Neural Networks modeling methods to predict the number of fractures. Neural networks had the best prediction followed by XG Boost and GBT methods. By being able to predict the injuries that need hospital admission, hospitals will be able to allocate resources more efficiently.

Future research can focus on how the hospital will use the model results/ predictions.

- How much time in advance does the hospital needs the results?
- What is the value-add for a particular hospital? For example, does the hospital need to know the variables that were the reason for a case to be a "Discharge" or "Predicted?"

The answers to these questions will help chose the best model and subset of variables for real-world implementations of systems. The next step of this research would be to work closely with the stakeholder, in this case the hospitals, to understand the requirements and rebuild the models to reflect their need. Once that is done, the model can be applied in real-time and results can be observed to see the positive changes. Also, better data-capture systems or techniques can help understand the scope for model improvements. More research needs to be done on how to use the data collected and findings to make a real-time system to run statistical models on real-time data.

REFERENCES

Alexander, B. H., Rivara, F. P., & Wolf, M. E. (1992). The cost and frequency of hospitalization for fall-related injuries in older adults. *American Journal of Public Health, 82*(7), 1020–1023.

Atkinson, E. J., Therneau, T. M., Melton, L. J., Camp, J. J., Achenbach, S. J., Amin, S., & Khosla, S. (2012). Assessing fracture risk using gradient boosting machine (GBM) models. *Journal of Bone and Mineral Research: The Official Journal of the American Society for Bone and Mineral Research, 27*(6), 67–72. http://doi.org/10.1002/jbmr.1577

Burns, E. B., Stevens, J. A., & Lee, R. L (2016). The direct costs of fatal and non-fatal falls among older adults—United States. *J Safety Res, 58,* 99–103.

Centers for Disease Control and Prevention, National Center for Injury Prevention and Control (2016, August 5). Web–based Injury Statistics Query and Reporting System (WISQARS) [Online]. https://www.cdc.gov/injury/wisqars/index.html

Chari, S., McRae, P., Varghese, P., Ferrar, K., & Haines, T. P. (2013). Predictors of fracture from falls reported in hospital and residential care facilities: a cross-sectional study. *BMJ Open, 3*(8), e002948. https://doi.org/10.1136/bmjopen-2013-002948

Chen, T., & Guestrin, C. (2016). XGBoost: A scalable tree boosting system. *Proceedings of the 22nd ACM SIGKDD International Conference on Knowledge Discovery and Data Mining, KDD '16, 785–794.* https://doi.org/ 10.1145/2939672.2939785

Dale, J., Green, J., Reid, F., & Glucksman, E. (1995). Primary care in the accident and emergency department: I. Prospective identification of patients. *BMJ, 311.* 423. https://doi.org/10.1136/bmj.311.7002.423

Deprey, S. M., Biedrzycki, L., & Klenz, K. (2017). Identifying characteristics and outcomes that are associated with fall-related fatalities: Multi-year retrospective summary of fall deaths in older adults from 2005–2012. *Injury Epidemiology, 4,* 21. http://doi.org/10.1186/s40621-017-0117-8

Dhargave, P., & Sendhilkumar, R. (2016). Prevalence of risk factors for falls among elderly people living in long-term care homes. *Journal of Clinical Gerontology & Geriatrics, 7,* 99–103. https://doi.org/10.1016/j.jcgg.2016.03.004

Eunsun, O., Seo, S.-W., Cheol Yoon, Y., Wook Kim, D., Kwon, S., Sungroh, & Yoon, S. (2017). Prediction of pathologic femoral fractures in patients with lung cancer using machine learning algorithms: Comparison of computed tomography-based radiological features with clinical features versus without clinical features. *Journal of Orthopaedic Surgery, 25*(2), 1–7. https://doi.org/10.1177/2309499017716243

Greenwood, D (1991). An overview of neural networks. *Behavioral Science, 36*(1), 1–33. http://dx.doi.org/10.1002/bs.3830360102

Ho-Le, T. P., Centerm J. R., Eisman, J. A., Nguyen, T. V., & Nguyen, H. T. (2017). Prediction of hip fracture in post-menopausal women using artificial neural network approach. Engineering in medicine and biology society (EMBC). *39th Annual International Conference of the IEEE, 4207–4210.* http://doi.org/10.1109/EMBC.2017.8037784

Ioannidis, G., Jantzi, M., Bucek, J., Adachi, J. D., Giangregorio, L., Hirdes, J., Pickard, L., & Papaioannou, A. (2017). Development and validation of the Fracture Risk Scale (FRS) that predicts fracture over a 1-year time period in institutionalized frail older people living in Canada: An electronic record-linked longitudinal cohort study. *BMJ Open, 7*(9), e016477. https://doi.org/10.1136/bmjopen-2017-016477

Jin, H., Lu, Y, Harris, S. T., Black, D. M., Stone, K. L., Hochberg, M. C., & Genant, H. K. (2004). Classification algorithms for hip fracture prediction based on recursive partitioning methods. *Medical Decision Making, 24*(4), 386–398. https://doi.org/10.1177/0272989x04267009

Kruse, C., Eiken, P., & Vestergaard, P. (2017). Machine learning principles can improve hip fracture prediction. *Calcified Tissue International, 100*(4), 348–360. https://doi.org/10.1007/s00223-017-0238-7

Lin, C. C., Ou, Y. K., Chen, S. H., Liu, Y. C., & Lin, J (2010). Comparison of artificial neural network and logistic regression models for predicting mortality in elderly patients with hip fracture. *Injury, 41*(8), 869–873. https://doi.org/10.1016/j.injury.2010.04.023

Menendez, M. E., & Ring, D. (2015). Factors associated with hospital admission for proximal humerus fracture. *American Journal of Emergency Medicine, 33*(2), 155–158. https://doi.org/10.1016/j.ajem.2014.10.045

Miedany, Y., El Gaafary, M., Toth, M., Palmer, D., Ahmed, I. (2011). Falls risk assessment score (FRAS): Time to rethink. *Journal of Clinical Gerontology & Geriatrics, 2*(1), 21–26. https://doi.org/10.1016/j.jcgg.2011.01.002

Miller, P. D., Barlas, S., Brenneman, S. K., Abbott, T, A., Chen, Y.-T., Barrett-Connor, E., & Siris, E, S. (2004). An approach to identifying osteopenic women at increased short-term risk of fracture. *JAMA Internal Medicine, 164*(10), 1113–1120. https://doi.org/1010.1001/archinte.164.10.1113

National Safety Council. (2003). *National Safety Council Injury Facts 2003 edition.* Retrieved October 2, 2019, from https://nfsi.org/nfsi-research/quick-facts/

Nilsson, J., Ohlsson, M., Thulin, L., Höglund, P., Nashef, S. A. M., & Brandt, J. (2006). Risk factor identification and mortality prediction in cardiac surgery using artificial neural networks. *Journal of Thoracic Cardiovascular Surgery, 132*(1), 12–9. e1. https://doi.org/10.1016/j.jtcvs.2005.12.055

Orces, C. H., & Alamgir, H. (2014). Trends in fall-related injuries among older adults treated in emergency departments in the USA Injury Prevention. *BMJ Journal, 20*, 421–423). http://doi.org/10.1136/injuryprev-2014-041268

Povalej Brzan, P., Obradovic Z., & Stiglic G. (2017). Contribution of temporal data to predictive performance in 30-day readmission of morbidly obese patients. *PeerJ* 5:e3230 https://doi.org/10.7717/peerj.3230

Shankar, K. N., Liu, S. W., & Ganz, D. A. (2017). Trends and characteristics of emergency department visits for fall-related injuries in older adults, 2003–2010. *Western Journal of Emergency Medicine, 18*(5), 785–793. http://doi.org/10.5811/westjem.2017.5.33615

Song, Y., & Lu, Y. (2015). Decision tree methods: applications for classification and prediction. *Shanghai Archives of Psychiatry, 27*(2), 130–135. http://doi.org/10.11919/j.issn.1002-0829.215044.

Sterling, D., A., O'Connor, J., A., & Bonadies, J. (2001). Geriatric falls: injury severity is high and disproportionate to mechanism. *Journal of Trauma–Injury, Infection and Critical Care, 50*(1), 116–119.

Stevens, J., & Sogolow, E. (2005). Gender differences for non-fatal unintentional fall related injuries among older adults. *Injury Prevention, 11*(2), 115–119. http://doi.org/10.1136/ip.2004.005835

Terrin, N, Schmid, C. H., Griffith, J. L., D'Agostino, R. B., & Selker, H. P. (2003). External validity of predictive models: a comparison of logistic regression, classification trees, and neural networks. *Journal of Clinical Epidemiology, 56*(8), 721–729. https://doi.org/10.1016/S0895-4356(03)00120-3

Verma, S. K., Willetts, J. L., Corns, H. L., Marucci-Wellman, H. R., Lombardi, D. A., & Courtney, T. K. (2016). Falls and fall-related injuries among community-dwelling adults in the United States. *PLoS One, 11*(3), e0150939.

Weycker, D., Edelsberg, J., Barron, R., Atwood, M., Oster, G., Crittenden, D. B., & Grauer, A. (2017). Predictors of near-term fracture in osteoporotic women aged ≥65 years, based on data from the study of osteoporotic fractures. *Osteoporosis International, 28*(9), 2565–2571. http://doi.org/10.1007/s00198-017-4103-3

World Health Organization. (2018). *Falls.* Retrieved September 2019, from https://www.who.int/news-room/fact-sheets/detail/falls

SECTION II

BUSINESS INTELLIGENCE AND OPTIMIZATION

CHAPTER 4

BUSINESS INTELLIGENCE AND THE MILLENNIALS

Data Driven Strategies for America's Largest Generation

Joel Thomas Asay, Gregory Smith, and Jamie Pawlieukwicz
Xavier University

ABSTRACT

In 2015, the census bureau projected the millennial generation would surpass the Baby Boom generation as the largest living demographic cohort. Millennials, those who have reached adulthood since the year 2000, represent as much as $1.4 trillion dollars in direct spending, and even more indirectly through the consumption behavior of their families. Even though peak millennial buying power is still several decades away, learning how to understand this generation presents opportunities to develop meaningful consumer relationships lasting long into the future. This chapter presents strategies for modifying current business intelligence (BI) practices to better suit the needs of millennial consumers. To accomplish this, we examine peculiarities of the millennial generation with the help of data from several

Contemporary Perspectives in Data Mining, Volume 4, pp. 41–51
Copyright © 2021 by Information Age Publishing
41

sources, including Xavier University's American Dream Composite Index. We focus on millennial perspectives on trust in businesses and government, spending behaviors, corporate social responsibility, methods of communication used among Millennials, and the shift from traditional data sources to next generation data mining.

INTRODUCTION

It is hard to pick up a business text or journal without seeing the buzzword "millennial." Viewed as the next big driver of the consumption economy, the "millennial" generation, or generation Y, is larger than any other demographic and arguably influences more dollars spent in the U.S. economy than any other cohort (Moos et al., 2018). Comprising the ages 18–34, millennials have grown up in an era of rapid technological advancement, and uncertain political environment, globalization and idealism—much like their "boom" generation parents. They also represent 30% of all retail sales (Berger, 2016). Millennials are more educated than any group before them and many graduated from college during a financial collapse, yet they exhibit financial behaviors unlike other generations (Moos et al., 2018). These defining characteristics and others are precisely why many marketing professionals and business strategists struggle to understand and capitalize on the millennials.

Like other disciplines, the BI approach needs to be attuned to capitalize on millennial consumers. This chapter presents a series of unique behaviors exhibited by members of the millennial generation that BI strategists should be familiar with, as well as a few tactics for how to customize your BI approach to better serve millennial customers. These strategies are designed to improve engagement and communication with millennials, prepare for the petabytes of new data they generate, and fine-tune data mining and analytical methods to take advantage of and predict millennial behavior.

The Importance of Making Adjustments Today

While the current spending influence of millennials today cannot be denied, it is only one factor to understanding their relative importance. The direct spending influence of Generation Y is estimated to be as high as $1.4 trillion dollars (Barton et al., 2014) and millennials already spend more money online annually than any other age range (Forrester Research, 2013). Obviously, millennial spending represents a significant part of the economy and while substantial, it only begins to scratch the surface of the broader impact they have.

Millennials *influence* an even larger sum of spending as they function in their role as "technology advisor" for the big financial discretionary decisions their family members make. Ask a boomer which laptop, tablet, or flat screen TV they bought and why, and chances are they will refer you to their son or daughter. The rebound phenomena many millennials have exhibited by moving back in with their parents after college only further influences the $3 trillion dollars in consumer spending the boomers exhibit, especially since many boomers paid for their millennial children's education in the first place (Bureau of Labor Statistics, 2013).

As much as millennials influence the purchasing behaviors of their families, they are highly influenced by the purchasing behavior of others. In a recent survey (ADCI, 2018), millennials were two to four times more likely than other generations to report that they feel pressured to make new purchases when they see others around them doing so. Social media and the modern ease of communication play a large part in millennials' propensity to look externally for defining themselves, and particularly their spending behavior.

Technology, in general, does well to define the millennial generation. Younger millennials may have had smartphones in grade school. Older millennials grew up expecting mail in a digital box before a physical one. More importantly for businesses, they report being significantly more likely to try *new* technology in comparison to other generations (ADCI, 2018).

How else do millennials differ from other cohorts? They are more educated than other generations, get married later in life, desire flexibility over home ownership, have more trust in the government and businesses, are more diverse and report being more inclusive of ideas that differ from their own (ADCI, 2018). Perhaps more importantly, they are more heterogeneous in their responses compared to other generations, even after accounting for location, income, and marital status.

Such differentiation from other generations makes it difficult to cater business decisions towards them. In the past, BI initiatives have provided large returns to understanding customer behavior, but with the millennial's diversity and the increase in the unstructured data they generate, adjustments to current BI initiatives are inevitable. As a result, the following strategies should be considered when interacting with a Generation Y customer.

ADAPTING DATA MINING AND ANALYTICS METHODS

Data Collection 2.0

The low hanging fruit in the world of data collection is structured data. Observable characteristics such as customer demographics, time, location, and purchasing behavior are easy to understand and analyze. These data sources are quantified and stored in data warehouses and databases where predefined models are easily applied. Contrast these data sources with unstructured sources such as customer service communication, social media interaction and customer specific behaviors. In truth, many companies in the past have avoided these types of data mining efforts because of their relative difficulty to effectively mine (Doan et al., 2008).

Beyond the mining of traditional unstructured data sets, which are often recorded as text rather than numerics, are next generation data opportunities that are particularly applicable to the millennial generation. For the purpose of this chapter, these next generation data sets can be broken into two primary groups—metadata and encoded digital data. Metadata is simply data that is used to describe or augment other data. These data can often be found attached to sources of the second category, encoded digital data. This includes images, digital voice recordings, and videos. As millennials migrate away from most text-based communication, the importance of mining the data they generate instead becomes even more important.

Metadata Necoming Megadata

The majority of the data we generate as consumers is done so without our even knowing it. Much of it is the result of our networked lives and our interconnectedness with technology. A common example is the data collected through behavior on the internet. When visiting a website, a customer may share data about themselves they never expected, such as the internet browser used, hardware, operating system, the size and resolution of the display used for viewing the website, approximate location, time spent on a page and much more. The implications to businesses operating an online marketplace are obvious, but some companies have even begun using this metadata to predict whether an individual will be a successful employee. For example, individuals who use internet browsers such as Google's Chrome browser or Mozilla's Firefox browser, tend to be higher performers in the workplace and experience less turnover ("Robot Recruiter," 2013). Another example of minable metadata is the information connected to new millennial communication methods, including images.

Instagram to Insta-Data

Millennials are continually breaking the norm for methods of communication. Companies, such as General Electric and Audi, are attracting millennial customers by employing Snapchat for communication with customers, rather than traditional e-mail or SMS which may be ignored by millennial customers (Carracher, 2015). Mercedes-Benz has reportedly had success using Instagram to communicate with and advertise to millennials (Instagram, 2015). With this said, what do Snapchat and Instagram have to do with data mining?

The Snapchat and Instagram success story goes beyond marketing. Because these are interactive methods of communication, they allow millennials to also share photos with businesses, and many do. While image-mining is only beginning to gain popularity in business realms, it holds significant potential as the tools are being developed to extract multiple forms of data from digital photographs (Foschi et al., 2002).

The first and more obvious form of data being collected from images can be found in the metadata attached to image files. Unbeknownst to many, information such as camera make and model, software, photograph settings, and even latitude and longitude can be captured by an image. This data can be particularly valuable because of the popularity of smartphones as cameras, and even knowing the type of smartphone used by your customers can be helpful in predicting behavior (O'Reilly & YouGov, 2014).

Second, and perhaps even more important, some companies are finding success using software to scan photos and automatically extract data such as brands, personal disposition and context (Macmillan & Dwoskin, 2014). Optical character recognition allow these tools to interpret what products people are mixing together and facial recognition can indicate sentiment. Is the subject enthusiastically happy, or frustrated holding your product? Background, scene, and GPS data can indicate where products are being used, what other activities consumers engage in when consuming the product or more generally, what your customer's other interests outside your product might be.

Image mining is a next generation method to collecting information and predicting customer behavior. Since millennials report being significantly more likely to include photography in their shopping behavior (Sprint, 2012), ensuring your company has a strategy to mine data from this new form of communication is even more important.

Prepare Your Analytics Team With the Best Tools Available—And Save Money Doing It

When computer assisted BI strategies were first gaining a foothold in American businesses, individual data points could be stored in just a few bytes for each customer. Descriptive data like gender and age require very little space, but new data sources such as images and customer metadata will greatly increase storage requirements. One of the largest social media companies maintains a 300-petabyte data warehouse and takes in about 600 terabytes per day for 1.39 billion users (Vagata & Wilfong, 2014).

Using traditional extract transform load methods require data to be moved from a relational database and into a shadow file system before statistical analysis is performed. The replication of data is expensive and requires additional space and infrastructure, as well as a considerable time penalty while data is queried, copied and transferred before any analysis can be performed (Brobst et al., 2009). Because patterns in millennial behavior are so varied, change quickly, and require large amounts of ware-housing space, this traditional method of analytics leaves a lot to be desired.

Instead of duplicating data in order to perform analysis, integrate an in-database processing tool within data warehouse. This approach includes native SQL functions for routine analysis, and can integrate with statisti-cal programming languages such as R and SAS (SAS Institute Inc., 2007). In its most simple form, such tools allow for the transformation of SAS commands into SQL and for the analysis that is too complicated to be completed in SQL, it allows for easy sandboxing of data into repositories where additional modeling can be completed. These tools can also be easily added to a NoSQL database where the handling of unstructured data is made much simpler (Leavitt, 2010). Privacy concerns surrounding the duplication and security of data can also be alleviated by incorporating some analytical tools directly into the database (Brobst et al., 2009).

Millennials as the Operators of Your BI Strategy

Combining in-database analytics with a real-time networking approach will allow models to run automatically and pushed to users as soon as they are available as opposed to waiting for users to query the system for updates. This is also in harmony with the way new millennial college graduates best respond to information. Among millennials, a push methodology is expected, not just helpful (Barton et al., 2014). The millennial drive to be connected to technology and updates almost always includes an expecta-tion that analytics be possible anywhere. Capitalize on this by employing tablet and smartphone platforms that extend the benefit through dash-boards and visual displays—both of which make data comprehension easier when making decisions quickly.

Millennials will constitute over 50% of the global workforce by 2020 (PricewaterhouseCoopers, 2011). These recent grads are better educated than any other cohort, and want to add their expertise to your customer analytics team. Provide them with the updated tools and strategies they work best with in order to make the most of them.

ENGAGING WITH MILLENNIAL CUSTOMERS

Return on Involvement: The New ROI

Millennials are unique in that they desire a relationship with a company beyond sale and consumption. They want to engage with and feel a sense of belonging to the companies they do business with (George, 2009). In practice, millennials need more than a traditional buyer/seller relationship to exhibit loyalty (Lazarevic, 2012). Thankfully, they are also more inclined to trust businesses more than other generations, so combine their willingness to trust and desire for involvement with their technological prowess.

A side-effect of better engagement with millennials is an increase in the data available for understanding them as a customer. Numerous companies maintain a Twitter or Facebook presence exclusively for the use of customer service interaction. While such a presence can obviously return value in the human resources aspect of operations, it is a mistake to not take advantage of social media's data. Several tools for mining text from Twitter and Facebook exist for popular statistical applications such as R. This social media data extends beyond the traditional customer service design and can identify trends in regional customer behavior, demand for new or revised products and services, comparisons with competitor products, or changes in the macroeconomic climate as it relates to your specific customer. This source of data also has a very low marginal cost of collection, since the data is already being generated and collected by the customer service division of your firm.

Social media also presents an opportunity to explore the *non*consumer behavior of your customers. While it may initially seem irrelevant, millennials in particular express a desire to be engaged with corporate social responsibility efforts and they report that it strongly influences their purchasing behavior (Cone Communications, 2013). Nationally, millennials express more interest in environmentalism and education, but this generation varies greatly in what social issues are important to them (ADCI, 2018). As mentioned earlier, Millennials are more heterogeneous than any other demographic construct. Millennials on the east and west coasts maintain more trust in their government, and Millennials in the rocky mountain west and mid-west report having significantly less trust in the government. While this may not be surprising given the political climate in these regions, even within a single location millennial respondents are almost twice as varied in their responses as other demographics (ADCI, 2018).

Avoid choosing corporate responsibility initiatives based on convenience and instead use social media mining to see what is important to your specific customer to earn their loyalty and business. Tactics such as these are even more important with millennials because research has indicated that 88% of Millennials report that they would switch brands if it would help a good cause. Over half of all millennials say they have even purchased a product because doing so benefited a cause, compared to 38% for all other adult respondents (Ferguson & Goldman, 2010).

Speaking of Loyalty, Your Loyalty Program Is Out of Date

Customer loyalty programs have provided numerous benefits to management over the years. One of the primary goals of loyalty program is increased understanding on customer specific buying behavior. Improving active adoption rates among millennials has been shown to be a function of "fun" and "inclusion" (Tews et al., 2015), and loyalty programs may provide such a feeling of inclusion when implemented appropriately (Nowak et al., 2006), and engaging smartphone applications can certainly be fun.

When it comes to examining effective strategies for engaging millennials in loyalty programs, it may be helpful to investigate the approaches employed by successful firms. Both Starbucks and Domino's Pizza have used engaging smartphone apps that require little commitment upfront, but provide loyalty programs and/or track shopping behavior easily. These programs secure customers with an easy to use interface and program that provides value to the customers by making rewards or incentives easy to understand and achievable. Because age, gender and location are sufficient to identify 87% of individuals (King, 2011), do not make the initial requirements overly complicated. Other perks, such as including a method of payment, can make the experience fun for millennials and reduce transaction costs.

This type of intuitive loyalty program has pushed Starbucks to become the most highly valued restaurant brand among millennials (Goldman Sachs, 2014). This is particularly curious since millennials are still several decades off from their peak earning potential, yet have chosen to venerate and patronize a luxury brand above all others. The most popular brand of smartphone among millennials is also the most expensive (Prosper Insights & Analytics, 2013), and luxury consuming behavior like $5 dollar coffee, technology and organic foods are now considered necessities by millennials.

Anchoring your loyalty program to price discounts, coupons or similar discounts as incentives is just another way of competing on price. Millennials have been said to be even more price sensitive than other consumers, but their unique behaviors also make them different. Millennials place a higher value on their social image (Fromm et al., 2011), are less satisfied with their present social status (ADCI, 2018) and "value the role they play in their communities" more so than other generations (The Council of Economic Advisers, 2014).

To capitalize on their desire for inclusion, consider using your loyalty program to appeal to millennials with more than pricing strategies. This might include membership in an exclusive group which rewards loyalty with perks that cannot be bought such as extended return policies, tailored customer service and early access or products or events. The millennial desire to earn rewards, benefits and inclusivity is exhibited in the fact that one third of millennials admit they have made a purchase they did not need just for the loyalty program perks (Bond Brand Loyalty, 2014).

In the digital age, the best advertisement for your program are the perspectives of the users voiced via social media and other communication outlets. Unfortunately for businesses, complaints and anger spread much more easily on social media than praise or joy (Fan et al., 2014). Gaining the loyalty of millennials may be a slow process, but losing trust and goodwill becomes easier in the digital age.

CONCLUSION

Despite growing up in uncertain economic and political times, the millennial generation is more optimistic about their potential for success and their role in society. They will define the next several decades of consumption, so ensuring a proper BI strategy is in place today will provide loyalty and data for years to come. Preparing for data storage requirements, adjusting data mining methods and analytics and improving the quality and type of communication between your business and millennials will go a long way to provide this data, and ultimately secure brand loyalty among America's largest generation of consumers.

REFERENCES

ADCI. (2018, April). *The American Dream Composite Index, Powered by Dunnhumby 2015 Report.* Xavier University's American Dream Composite Index. www.xavier.edu/adci

Barton, C., Koslow, L., & Beauchamp, C. (2014). *The reciprocity principle: How millennials are changing the face of marketing forever.* The Boston Consulting Group.

Berger, A. (2016). *Marketing and American consumer culture.* Palgrave Macmillan.

Bond Brand Loyalty. (2014). *The 2014 Loyalty Report.* Bond Brand Loyalty.

Brobst, S., Collins, K., Kent, P., & Watzke, M. (2009). *High performance analytics with in-database processing.* Terradata. Retrieved April 25, 2015, from http://analytics.ncsu.edu/sesug/2009/AD001.Brobst.Kent.pdf

Bureau of Labor Statistics. (2013). *Consumer Expenditure Survey*. United States Department of Labor.

Carracher, J. (2015). *How 5 technology brands are crushing it on snapchat*. Consumer Electronics Association. Retrieved April 5, 2015, from http://www.ce.org/Blog/Articles/2015/March/How-5-Technology-Brands-are-Crushing-it-on-Snapcha.aspx

Cone Communications. (2013). *2013 Cone communications social impact study: The next cause evolution*.

Council of Economic Advisers. (2014). *15 economic facts about millennials*. Executive Office of the President of the United States. Retrieved March 10, 2015, from https://www.whitehouse.gov/sites/default/files/docs/millennials_report.pdf

Doan, A., Naughton, J. F., Ramakrishnan, R., Baid, A., Chai, X., Chen, F., Chen, T., Chu, E., DeRose, P., Gao, B., Gokhale, C., Huang, J., Shen, W., & Vuong, B.-Q. (2008). Information extraction challenges in managing unstructured data. *ACM SIGMOD Record, 37*(4), 14–20.

Fan, R., Zhao, J., Chen, Y., & Xu, K. (2014). Anger is more influential than joy: Sentiment correlation in Weibo. *PLoS ONE, 9*(10).

Ferguson, R., & Goldman, S. (2010). The cause manifesto. *The Journal of Consumer Marketing, 27*(3), 283–287.

Forrester Research. (2013). *North American technographics Online Benchmark Survey*. Cambridge. https://www.forrester.com/North+American+Technographics+Online+Benchmark+Survey+Part+1+2013/-/E-sus2111#

Foschi, P. G., Kolippakkam, D., Liu, H., & Mandvikar, A. (2002). Feature extraction for image mining. *Multimedia Information Systems*, 103–109.

Fromm, J., Lindell, C., & Decker, L. (2011). *American millennials: Deciphering the enigma generation*. Barkley.

George, B. (2009, Oct. 6). *Brands are dead. Welcome to the paricipation economy*. Retrieved March 10, 2015, from http://www.billgeorge.org/page/kevin-roberts-brands-are-dead--welcome-to-the-participation-economy

Goldman Sachs. (2014). *Goldman Sachs survey of 2,000 US customers*. Author.

Instagram. (2015). *A Compact SUV for the millennial lifestyle*. Instagram for business. Retrieved April 15, 2015, from http://instagram-static.s3.amazonaws.com/MB_CaseStudy_FINAL1.pdf

King, G. (2011). Ensuring the data-rich future of the social sciences. *Science, 13*, 720.

Lazarevic, V. (2012). Encouraging brand loyalty in fickle generation Y consumers. *Young Consumers, 13*(1), 45–61.

Leavitt, N. (2010). Will NoSQL databases live up to their promise? *Computer, 43*(2), 12–14.

Macmillan, D., & Dwoskin, E. (2014, Oct. 9). Smile! Marketing firms are mining your selfies. *The Wall Street Journal*. Retrieved April 8, 2015, from http://www.wsj.com/articles/smile-marketing-firms-are-mining-your-selfies-1412882222

Moos, M., Pfeiffer, D., & Vinodari, T. (2018). *The millennial city*. Routledge.

Nowak, L., Thach, L., & Olsen, J. (2006). Wowing the millennials: Creating brand equity in the wine industry. *The Journal of Product and Brand Management, 15*(5), 316–323.

O'Reilly, L., & YouGov. (2014). *Apple people and Android people even eat different foods.* Business Insider. https://www.businessinsider.com.au/apple-users-vs-android-users-according-to-yougov-profiles-2014-11

PricewaterhouseCoopers. (2011). *Millennials at work reshaping the workplace.* PwC.

Prosper Insights & Analytics. (2013). *Hot: Samsung Galaxy S4 versus Apple iPhone 5 by generation.* Retrieved March 15, 2015, from http://prosperdiscovery.com/though-android-continues-to-climb-millennials-are-looking-to-apple/

Robot recruiters. (2013, April 6). The Economist. Retrieved April 15, 2015, from http://www.economist.com/news/business/21575820-how-software-helps-firms-hire-workers-more-efficiently-robot-recruiters?fsrc=scn/tw_ec/robot_recruiters

SAS Institute Inc. (2007). *SAS In-Database processing: A roadmap for deeper technical integration with database management systems.* Retrieved April 10, 2015, from https://support.sas.com/resources/papers/InDatabase07.pdf

Sprint. (2012). *Mobile moment of truth survey.* Author.

Tews, M., Michel, J., Xu, S., & Drost, A. (2015). Workplace fun matters ... but what else? *Employee Relations, 37*(2), 248–267.

Vagata, P., & Wilfong, K. (2014). *Scaling the Facebook data warehouse to 300 PB.* Facebook. Retrieved Mar 10, 2015, from https://code.facebook.com/posts/229861827208629/scaling-the-facebook-data-warehouse-to-300-pb/

CHAPTER 5

DATA-DRIVEN PORTFOLIO OPTIMIZATION WITH DRAWDOWN CONSTRAINTS USING MACHINE LEARNING

Meng-Chen Hsieh
Rider University

ABSTRACT

In practice, data-driven optimal portfolio decisions are derived based on the time series data of target asset returns. Such data-driven optimization decision rules are prone to inferior out-of-sample performance due to estimation errors of parameters plugged in the optimization setting. In the "big data" era, correlations between target asset returns and auxiliary variables are frequently observed. These auxiliary variables have the potential to provide valuable information on their association with the target asset returns and thus may be able to improve the out-of-sample performance of the constructed optimal portfolio. In this work, we consider a portfolio optimization problem with drawdown constraints. We apply machine learning methods to leverage the association between target asset returns and auxiliary variables to derive optimal portfolio decisions. A comparison study on the out-of-sample performance of the constructed portfolio with and without

Contemporary Perspectives in Data Mining, Volume 4, pp. 53–75
Copyright © 2021 by Information Age Publishing

utilizing machine learning methods shows the improvement of implementing machine learning in optimal portfolio decisions.

INTRODUCTION

A portfolio manager's skill is measured by the returns of his managed fund
as well as the risks underlying the returns. A common risk measure is the
volatility of the returns which quantifies the average deviation of returns
from the mean. A fund with a large average return and a higher volatility is
less favorable than a fund with a larger average return but a lower volatility,
since a higher volatility implies a less stable rate of returns. The Markowitz
minimum variance portfolio is a classical portfolio optimization problem
in which portfolio weights are derived such that given a target portfolio
return, the variance of the portfolio returns is minimized. In addition to
the volatility as a risk measure, there are other measures such as value-at-
risk (VaR) and drawdown, both of which focus on the downside risk of a
portfolio. For example, the VaR provides the expected loss of a portfolio
within a time period given a certain probability. A drawdown measures the
gap between the current portfolio cumulative return and the maximum
of up-to-date cumulative returns. A large drawdown indicates a decline in
the cumulative return of the fund and thus has a negative impact on an
investor's accumulated wealth. To protect the money invested in a fund,
investors usually require a constraint on the value of drawdown. If a fund's
return reaches the drawdown constraint, investors are likely to withdraw
their money from the fund. Since a significant portion of a portfolio manager's compensation comes from the management fees as a percentage
of money invested in the fund, from the portfolio manager's perspective,
there is a strong incentive to manage the drawdown under a certain level
while maximizing the returns of the fund. Therefore, optimal portfolio
decisions on minimizing drawdown are widely used in asset management
industry, and have drawn attention from academia. Grossman and Zhou
(1993) and Cvitanic and Karatzas (1995) derive analytical solutions for
optimal portfolio decisions on minimizing drawdown. Chekhlov et al.
(2003, 2005) propose the *CDaR* (Conditional Drawdown at Risk) approach
to measure the drawdown risk and apply a linear programming method to
prescribe optimal portfolios with drawdown constraints.

In practice, these optimal portfolio decision rules are derived based
on the historical data of the underlying assets. For example, to derive the
Markowitz minimum variance portfolio, one first estimates the means and
the covariances of assets returns using the historical return data, and then
plugs in these statistical estimates to derive the optimal portfolio weights.
In the case where the optimal portfolio seeks to maximize the average

returns with a constraint on the drawdown, one relies on historic returns to quantify the portfolio drawdown. Therefore, the derived portfolio weights in these portfolio optimization problems are data driven and prone to have inferior performance during the out-of-sample period. For example, Ban et al. (2018) show that portfolio optimization model has limited impact in practice because of estimation issues when applied to real data. They propose a performance-based regularization approach to reduce the impact of estimation errors when deriving the optimal portfolio decision.

In the era of "big data," information dissipation in financial market comes from a variety of channels. It may be directly from the real-time financial market data, internet traffic, or online activities such as the release of news on the social media and websites. Therefore, portfolio decisions may not solely depend on historical return data. They may also depend on variables from other resources. Bertsimas and Kallous (2020) introduce a general framework of applying machine learning methods to prescribe data-driven decision rules. In their illustrative examples, Bertsimas and Kallus (2020) show that the data-driven decision rules utilizing machine learning methods generate a better performance in the validation data than those without utilizing machine learning methods. This work builds upon the approach proposed in Bertsimas and Kallus (2020) by presenting a detailed procedure on applying machine learning methods to a portfolio optimization problem with drawdown constraints.

Our contributions are twofold: first, we show that the portfolio optimization problem with drawdown constraints can be reduced into a linear programming problem. Second, we establish a link connecting machine learning methods with the linear programming problem to prescribe a data-driven optimal portfolio decision. A simulation-based comparison study on the out-of-sample performance of the constructed portfolio with and without utilizing machine learning methods shows the improvement of implementing machine learning in optimal portfolio decisions. Practitioners can benefit from applying this procedure for constructing an optimal portfolio using real-world data.

The chapter is organized as follows: in the next section, we review a general framework introduced by Bertsimas and Kallous (2020) of applying machine learning methods to prescribe optimal decision rules under uncertainty. We then define a portfolio optimization problem with drawdown constraints. We show that the portfolio optimization problem can be reduced into a linear programming problem. We combine machine learning and the linear programming problem to prescribe the data-driven optimal portfolio decision. Finally, we present a simulation study and evaluate the out-of-sample performance of the constructed portfolio utilizing machine learning methods. We conclude the work in the Conclusion section.

APPLICATION OF PREDICTIVE ANALYTICS TO
PRESCRIBE OPTIMAL DECISIONS

The optimal portfolio decision problem is one instance of the stochastic optimization problems where the objective function involves variables with uncertainty. The traditional decision rule making under uncertainty is presented as

$$z \in \arg\min_{z \in \mathcal{Z}} E[v(z;Y)] \, , \tag{1}$$

where $v(z;Y)$ is the objective function with random inputs $Y \in \mathbf{R}^{dy}$ and z is the derived optimal decision. A popular data-driven approach to solve the stochastic optimization problem in (1) is the sample approximation average (SAA) which relies on the past observations of Y without the inputs of the auxiliary variables:

$$\hat{z}_n^{SAA} \in \arg\min_{z \in \mathcal{Z}} \frac{1}{n} \sum_{i=1}^{n} v(z;y_i) \, , \tag{2}$$

where $\{y_1,...,y_n\}$ are the sample realizations of Y.

In the past decade, with explosions in the availability and accessibility of data, people now have access to both the target variables $Y \in \mathbf{R}^{dy}$ as well as their concurrent auxiliary variables $X \in \mathbf{R}^{dx}$. It is quite common that one obtains pairs of concurrent target variables and auxiliary variables (y_1, x_1), (y_2, x_2), ..., (y_n, x_n) in the collected data. These auxiliary variables may contain information on their associations with the target variable and together they form a joint distribution. Bertsimas and Kallus (2020) propose a framework leveraging the association between the auxiliary variables and the target variables to derive the covariate-dependent optimal decision rule:

$$z^*(x) \in \mathcal{Z}^* \in \arg\min_{z \in \mathcal{Z}} E[v(z;Y)|X = x] \tag{3}$$

The solution $z^*(x)$ to (3) utilizes the joint distribution of (X,Y) and leverages the observation of $X = x$ to obtain the optimal decision $z^*(x)$. The idea behind Bertsimas and Kallus's (2020) approach is similar to the application of a non-parametric regression model to predict the target variable, where one computes the average value of ys in the nearby locations defined by a kernel function to approximate the conditional mean of the target variable. To prescribe the optimization rule given by (3), one first classifies the observed pairs of (y_1, x_1), (y_2, x_2), ..., (y_n, x_n) into several neighborhoods and evaluate the average value of the objective function in the nearby locations. The optimal decision rule is then derived by optimizing the local average of the objective function. Bertsimas and Kallus (2020) propose a data-drive decision rule for (3) (see Eq. (3) on Bertsimas and Kallus (2020) which takes the following general form

$$\hat{z}_n(x) \in Z^* \in \arg\min_{z \in Z} \sum_{i=1}^{n} w_{n,i}(x) v(z; y_i) ,$$

(4)

where $\omega_{n,i}(x)$ are the weights derived from the data:

$$w_{n,i}(x) = \begin{cases} 0 & : \quad (x_i, y_i) \notin N_k \\ \frac{1}{\sum_{i=1}^{n} 1_{\{(x_i, y_i) \in N_k\}}} & : \quad (x_i, y_i) \in N_k \end{cases}$$

(5)

Note the location set N_k is determined by the pairwise observations of x_i and y_i. A special case of $\omega_{n,i}(x)$ in (5) would be to treat each pair of (x_i, y_i) equally regardless of the local neighborhood N_k where (x_i, y_i) lies. In such a case, each pair receives an equal weight of $\frac{1}{N}$, where N is the total number of pairs in the data. Under the framework of (5), the prescribed decision rule $z^*(x)$ is dependent on the values of the auxiliary variables. Thus, rather than obtaining one single decision rule prescribed by (1), there might be multiple covariate-dependent decision rules prescribed by (3).

For data sets characterized by high dimensionality and a large number of observations, machine learning methods provide powerful tools to tackle the complicated structure between X and Y and thus are widely used in predictive analytics. In the following, we review several machine learning methods widely used in predictive analytics and illustrate their applications for prescribing optimal decision rules under the framework (3).

Review of Machine Learning Methods and Their Applications to Data-Driven Decisions

kNN(k-Nearest Neighbors)

The *k-Nearest Neighbors regression* (kNN) first identifies the k points in the training data that are closest to a given point x_0. It then estimates the average value of the response variable associated with these k points. We apply the idea from the *k-Nearest Neighbors regression* and derive the optimal decision rule conditional on the observations of the auxiliary variables in the local neighborhood. Let $v(z; y^i)$ represent the objective function dependent on the decision rule z and the observed inputs y^i. We prescribe our optimal decision rule

$$\hat{z}_N^{kNN}(x_0) \in \arg\min_{z \in \mathcal{Z}} \sum_{i \in \mathcal{N}_k(x_0)} v(z; y^i) \tag{6}$$

where

$$\mathcal{N}_k(x_0) = \left\{ i = 1, \ldots, N : \sum_{j=1}^N I[\|x_0 - x_i\| \geq \|x_0 - x_j\|] \leq k \right\}$$

is the neighborhood of the k data points that are closest to x_0. The value of k is the tuning parameter which determines how many points are included to estimate the average value of the response variable. The larger the value of k, the more points are included, and thus the method becomes more flexible.

Ctree (Conditional Inference Trees)

The *Conditional Inference Trees* (Ctree) proposed by Hothorn et al (2006) is an improvement of the *Classification and Regression Tree* (CART). The CART uses recursive binary splitting by performing an exhaustive search over all possible splits and selecting those splits maximizing an information measure of node impurity. This approach has two problems: overfitting and a selection bias towards covariates with many possible splits. To overcome these shortcomings, Hothorn et al (2006) introduced statistical inference tests to distinguish between a significant and an insignificant improvement in the information measure under a possible split. Hence the p–value of the statistical inference test determines how stringent the criteria for a split. A smaller p–value implies a more stringent criteria for a split and less flexible the method. We incorporate the Ctree method to prescribe the optimal decision rule

$$\hat{z}_N^{Ctree}(x) \in \arg\min_{z \in \mathcal{Z}} \sum_{i:R(x^i)=R(x)} v(z; y^i) \tag{7}$$

where $R(x)$ represents the splitting rule implied by a regression tree trained on the observed data.

Random Forests

Motivated by the CART and the Bootstrap Aggregation Trees (Bagging), Breiman's (2001) Random Forests build a number of decision trees based on the bootstrapped training samples. Each time when a split in a tree is considered, a random sample of m predictors is chosen as split candidates from a full set of p predictors. A random selection of the m predictors helps to decrease the correlations among the trees and thus makes the average of the resulting trees less variable and more reliable. The decision rule based on the random forests is proposed as follows

$$\hat{z}_t^{RF}(x) \in \underset{z \in \mathcal{Z}}{\arg\min} \sum_{i: R^t(x^i)=R^t(x)} v(z; y^i) \tag{8}$$

$$\hat{z}_N^{RF}(x) = \frac{1}{T_r} \sum_{t=1}^{T_r} \hat{z}_t^{RF}(x) \tag{9}$$

where $R^t(x)$ is the splitting rule implied by the tth tree in a random forest, T_r is the total number of trees grown in the forest. Equation (8) is essentially the same rule carried out by the Ctree method for a single tree. Once we obtain the decision rule for each tree in a forest, we take the average of the rules across all trees.

When carrying out the random forest method, the choices of the tuning parameters play an essential role in the prediction/classification results and thus have an impact on the final optimal decision rule. For example, a typical choice of the m predictors is $m = \sqrt{p}$. When a large number of predictors are correlated, selecting a small number of predictors is very effective. Second, the choice of p –value of the statistical inference for splitting a tree. As mentioned in the previous section, the p –value determines how stringent a criterion is for a split. Hence the criterion imposed on a single tree will ultimately affects the outcome of an entire forest. Finally, the choice of the number of trees built in a forest. The more trees are built, the more deductions in the variability of the predictions.

In the next section, we present a portfolio optimization problem with a constraint in drawdown and illustrate the application of the described machine learning methods to derive data-driven optimal portfolio decisions.

PORTFOLIO OPTIMIZATION WITH DRAWDOWN CONSTRAINTS

Definition of Drawdown

Let $V(t)$ represent the value of a portfolio at time t. A portfolio's drawdown $D(t)$ is defined as

$$D(t) = \left\{ \max_{0 \leq \tau \leq t} V(\tau) \right\} - V(t) . \tag{10}$$

$D(t)$ is a time series which value is determined by the difference between the portfolio value and its up-to-date maximum at time t. Figure 5.1 provides an illustration of the dynamics of $D(t)$. Since $D(t)$ reflects the drop of the portfolio value compared to its peak value, $D(t)$ can never be negative. A zero $D(t)$ indicates that the current portfolio value is also its peak value.

Optimal Portfolio With Drawdown Constraints

Assume there are m assets available for constructing an investment portfolio. Let j denote the asset index, where $j = 1, 2, 3, \ldots, m$, and $r_j(k)$ the return of the jth asset at time k. For each asset j, we observe a sample-path of its returns, $(r_j(1), r_j(2), \ldots, r_j(T))$. Based on the sample path of returns, we can compute the uncompounded cumulative return for the jth asset up to time t, $y_j(t) = \Sigma_{k=1}^t r_j(k)$. Let $y(t) = (y_1(t), y_2(t), \ldots, y_m(t))$ be a vector of the uncompounded cumulative returns of the m assets at t and ω_j be the weight the portfolio manager invests in asset j. The portfolio value at t is measured as the weighted average of the uncompounded cumulative returns from the m assets:

$$V(t) = \sum_{j=1}^m \omega_j y_j(t) = \omega^T y(t) \equiv V(\omega, y(t)) ,$$

where $\omega = (\omega_1, \omega_2, \ldots, \omega_m)$ is a set of investment weights in the m assets. Note since the drawdown process $D(t)$ is derived from $V(t)$, $D(t)$ also depends on the investment weights ω and the observed cumulative returns $y(t)$.

In this chapter, we focus on two types of drawdown constraints: (1) maximum drawdown $M(t) = \{\max_{0 \leq \tau \leq t} D(\tau)\}$ and (2) conditional draw-down-at-risk (CDaR) proposed by Chekholv et al. (2003, 2005). Given a confidence level $\alpha \in (0,1)$, the $CDaR_\alpha$ is defined as the average of the worst $(1 - \alpha) \times 100\%$ of the drawdowns measured over the interval $[0, T]$. Let ξ represent the minimum threshold value such that approximately $(1 - \alpha) \times$

100% of the drawdowns are greater than ξ. Chekhlov et al. (2003) propose the following general formula for an approximation of $CDaR_\alpha$:

$$CDaR_\alpha = \min_\xi \left\{ \xi + \frac{1}{(1-\alpha)T} \sum_{t=1}^{T} [D(t) - \xi]^+ \right\} \qquad (11)$$

When $\alpha = 1$, the $CDaR_\alpha$ is equal to the maximum drawdown since 0% of all drawdown will be larger then ξ. Thus the maximum drawdown is a special case of $CDaR_\alpha$. When $\alpha = 0$, the $CDaR_\alpha$ is essentially the average of all drawdowns. Therefore, when $0 < \alpha < 1$, the $CDaR_\alpha$ is between the average and the maximum of all drawdowns.

A portfolio manager faces a data-driven optimal portfolio problem may consider a constraint in $M(t)$ or the $CDaR_\alpha$. Specifically, given a sample path of the portfolio returns over the period $[0, T]$, a portfolio manager's portfolio optimization problem with *maximum drawdown constraints* is set up as

$$\arg\max_{\omega \in W} \frac{1}{T} V(\omega; y(T))$$

$$s.t. \qquad \left\{ \max_{0 \leq t \leq T} D(\omega, y(t)) \right\} \leq C_1 \qquad (12)$$

$$L \leq \omega^T 1 \leq 1.$$

The portfolio optimization with the $CDaR_\alpha$ constraint is set up as

$$\arg\max_{\omega \in W, \xi} \frac{1}{T} V(\omega; y(T))$$

$$s.t. \qquad CDaR_\alpha(\omega, y(t)) \leq C_2 \qquad (13)$$

$$L \leq \omega^T 1 \leq 1.$$

where C_1 and C_2 are constants. L is the lower bound for the sum of the investment weights. The constraint on the sum of the investment weights, $L \leq \omega^T 1 \leq 1$, ensures that the portfolio manager constructs the portfolio with the money invested in the fund. Note that this constraint may be changed due to the preference of the portfolio manager or the investors. For example, in addition to the constraint on the sum of the weights, there are constraints for each weight of the underling asset. The objective function, $\frac{1}{T} V(\omega, y(T))$, represents the average uncomponded cumulative return of the portfolio over the interval $[0, T]$. In the case of portfolio optimization with $CDaR_\alpha$ constraint, the decisions rule involves both the optimal portfolio weights and the minimum threshold ξ.

Linear Programming Problem

Chekholv et al. (2003, 2005) show that the portfolio optimization problem with drawdown constraint such as (12) or (13) is a convex optimization problem with a linear return function and piece-wise linear convex constraints. Therefore, the portfolio optimization problem can be reduced to a linear programming problem. For example, the portfolio optimization with a constraint in the maximum drawdown can be reformulated as a linear programming problem:

$$\arg\max_{\omega,u} \frac{1}{T} V(\omega; y(T))$$

$$s.t. \quad u_t - \omega^T y(t) \leq C \ , \ 1 \leq t \leq T$$

$$u_t \geq \omega^T y(t)$$

$$u_t \geq u_{t-1}$$

$$u_0 = 0$$

$$L \leq \omega^T 1 \leq 1 . \tag{14}$$

where u_t is the auxiliary variable implemented for solving the linear programming problem, for all $t = 1, 2, ..., T$.

In the case where there is a constraint on the *CDaR*, the liner programming formation is formulated as

$$\arg\max_{\omega,u,z,\xi} \frac{1}{T} V(\omega; y(T))$$

$$s.t. \quad \xi + \frac{1}{(1-\alpha)T} \sum_{t=1}^{T} z_t \leq C \ , \ 1 \leq t \leq T$$

$$z_t \geq u_t - \omega^T y(t) - \xi$$

$$z_t \geq 0$$

$$u_t \geq \omega^T y(t)$$

$$u_t \geq u_{t-1}$$

$$u_0 = 0$$

$$L \leq \omega^T 1 \leq 1 , \tag{15}$$

where both z_t and u_t are the auxiliary variables implemented for solving the linear programming problem, where $t = 1, 2, ... , T$.

Optimal Portfolio Construction Utilizing Machine Learning

We can extend the above portfolio optimization problem to a data-driven optimal portfolio problem utilizing the machine learning approaches as discussed in the previous review of machine learning methods section. For a given machine learning method, we first classify the observed data into several groups based on the characteristics of the covariates. We then solve the linear programming problem for each group of the data and derive the optimal investment weights. Therefore, rather than one single set of weights ω prescribed for portfolio construction, a multiple set of weights $\omega(x)$ are prescribed based on the observed covariates x. Specifically, the covariate dependent prescription for the optimal portfolio with maximum drawdown constraint is formulated as the following:

$$\underset{\omega(x)\in W,u}{\arg\max} \frac{1}{\mathcal{N}_k(x)} \sum_{t\in\mathcal{N}_k(x)} V\left(\omega(x); y(t)\right)$$

$$s.t. \quad u_t - \omega^T(x)y(t) \leq C \ , \ t \in \mathcal{N}_k(x)$$

$$u_t \geq \omega^T(x)y(t)$$

$$u_t \geq u_{t-1}$$

$$u_0 = 0$$

$$L \leq \omega^T 1 \leq 1 \ . \tag{16}$$

The prescription for the case with a constraint on the *CDaR* is formulated as:

$$\underset{\omega(x)\in W,u,z,\xi}{\arg\max} \frac{1}{\mathcal{N}_k(x)} V\left(\omega(x); y(t)\right)$$

$$s.t. \quad \xi + \frac{1}{(1-\alpha)\mathcal{N}_k(x)} \sum_{t\in\mathcal{N}_k(x)} z_t \leq C \ , \ t \in \mathcal{N}_k(x)$$

$$z_t \geq u_t - \omega^T(x)y(t) - \xi$$

$$z_t \geq 0$$

$$u_t \geq \omega^T(x)y(t)$$

$$u_t \geq u_{t-1}$$

$$u_0 = 0$$

$$L \leq \omega^T 1 \leq 1 \ . \tag{17}$$

SIMULATIONS

Simulation of Returns

In this section, we conduct a simulation study to demonstrate the application of machine learning to prescribe solutions for the optimal portfolio problems given by (16) and (17). We follow the same procedure as Bertsimas and Kallus (2020) for simulating the return process. We assume that there are 12 assets available for portfolio construction. Let $R_i(t)$ represent the return of asset i at t, where $i = 1, 2, 3, ..., 12$. For simplicity, we assume that the returns are driven by three factors denoted by $F(t) = \{F_1(t), F_2(t), F_3(t)\}$. Let A_i and B_i be the loadings of the three factors on the mean and variances of $R_i(t)$, and δ_i and η_i are *iid* standard Gaussian noises representing the non-systematic risks to asset i. The dynamics of $R_i(t)$ is generated by the following equation:

$$R_i(t) = A_i^T(F(t) + \delta_i/4) + (B_i^T F(t))\eta_i \ \forall i = 1, 2, \ldots, 12 \tag{18}$$

Thus $R_i(t)$ has mean $A_i^T(F(t))$ and variance $(B_i^T F(t))^2$. We assume that there are correlations among the three factors, which dynamics is governed by a 3-dimensional multivariate ARMA(2, 2) process

$$F(t) - \Phi_1 F(t-1) - \Phi_2 F(t-2) = U(t) - \Theta_1 U(t-1) - \Theta_2 U(t-2) , \tag{19}$$

where $U \sim N(0, \Sigma_U)$ are multivariate white noise processes. Simulations of $F(t)$ and specifications of A, B, U are provided in Appendix.

Simulation Study

We conduct a simulation study to evaluate performances of portfolios constructed from utilizing the machine learning methods. In each of the 100 realizations, we simulate the training data which contains concurrent pairs of returns and factors, $((R(t), F(t))$, $t = 1, 2, ..., T$. We consider six sizes of the training data, $T = 10, 100, 1000, 2500, 5000, 10000$, and a fixed size of 1000 observations in the validation set used for out-of-sample performance assessment. For a given size of the training data, we train a machine learning method with the training data and derive the optimal portfolio weights. We then apply the prescribed covariate-dependent portfolio weights to the validation set and evaluate the performance of the constructed portfolio.

Note that the choices of tuning parameters in implementing a machine learning method play an essential role in the classification outcomes. In the kNN method, the tuning parameter k is the number of the observations included in a neighborhood. We choose $k = T^{0.4}$, where T is the size of the training data, to classify the data into multiple groups and derive the optimal portfolio weights.

In the Ctree method, the level of significance $\alpha = 0.05$ is chosen to determine whether a split is statistically significant. For the Random Forest, its tuning parameters include the number of trees grown in a forest and a random number of m predictors selected from all candidate predictors.

We consider growing 10 trees in a forest and a random selection of $m = \sqrt{3}$ predictors from all 3 predictors in our study.

In both portfolio optimization problems, we use $C = 1.0$ for the constraints of maximum drawdown and $CDaR_\alpha$ respectively, and $L = 0.5$ for the lower bound of total portfolio weights in (16) and (17). For $CDaR_\alpha$, we set $\alpha = 0.05$. Hence we restrict the average of the worst 95% drawdowns of a sample path to be no more than 1.0, or 100%.

To compare the performance of portfolio constructed with and without utilizing machine learning methods, we consider three types of performance metrics: (1) **average cumulative return**, (2) **maximum drawdown**, and (3) **reward-risk ratio**. The reward-risk ratio is defined as the ratio of the average cumulative return to the maximum drawdown. Since a portfolio with high returns tends to be associated with high risks, we use the reward-risk ratio to measure the average cumulative return per unit of risk. We also consider a benchmark portfolio constructed purely based on the historical return process $R(t)$ without the inputs of the auxiliary variables. We denote such a benchmark portfolio as the "SAA" portfolio.

Figures 5.2 to 5.4 show the average out-of-sample performance metrics for portfolios with maximum drawdown constraints over the 100 realizations. Figure 5.2 shows the average cumulative returns over the 100 realizations for each method. It indicates that the benchmark portfolio based on the SAA approach, with almost flat cumulative returns, does not deliver a better out-of-sample performance than those utilizing the machine learning methods across all training sizes. The out-of-sample performances of portfolios utilizing machine learning methods demonstrate an improvement in the average portfolio cumulative return as the training size increases. Among the three machine learning methods, the Random Forest and the Ctree methods have a similar performance and both outperform the kNN method as T becomes larger. Figure 5.3 shows the average maximum drawdowns over the 100 realizations for each method. Among all methods, the Random Forest achieves the lowest average maximum drawdown. The asymptotic size effect starts kicking in as T increases for the SAA-based portfolio, as it gradually lowers its average maximum drawdown as the training size increases. Figure 5.4 shows the average out-of-sample risk-adjusted returns over the 100 realizations. The portfolio utilizing the Random Forest approach consistently has the highest average reward-risk ratio compared with the other approaches across all sizes. The reward-risk ratios of the portfolios utilizing machine learning methods outperform the ratio from the benchmark portfolio across all sizes.

Figures 5.5 to 5.7 show the distributions of performance metrics over the 100 realizations for each method. Note that in Figure 5.6, the distribution of maximum drawdown from the Random Forest based portfolio has the smallest variation of maximum drawdown compared with other methods. This suggests that the Random Forest method is more effective in meeting the maximum drawdown constraint. In terms of the variations of the other performance metrics such as the cumulative returns and reward-risk measure, the Random Forest as well as the Ctree and kNN methods, the variations of these two performance metrics are larger than the variations of those metrics from the SAA method. This can be explained by the smaller number of observations available for prescribing a decision rule after each split of data when running the machine learning methods.

Figures 5.8 to 5.13 show the comparisons of the average as well as the variations of the performance metrics for portfolios with constraints in the $CDaR_{\alpha = 0.05}$. Except for Figures 5.9 and 5.12, we find very similar patterns in these figures with those in the figures with portfolio constraints in the maximum drawdown. Specifically, we notice that in Figure 5.9, the SAA has a higher maximum drawdown than the machine learning methods across most sizes and the variation of the maximum drawdowns is getting bigger than the case with a constraint in the maximum drawdown. On the contrary, the machine learning methods, show a significant reduction in the variations of maximum drawdowns in Figure 5.12. When $\alpha = 0.05$, the $CDaR_{\alpha = 0.05}$ is very close to the average of drawdowns. Therefore, a constraint in the $CDaR_{\alpha = 0.05}$ helps to smooth out the variations in the drawdowns and thus reduces the variations of maximum drawdowns as well.

In summary, our simulation study shows that portfolios utilizing machine learning methods outperform the benchmark portfolios without utilizing machine learning methods in the validation data. In particular, the Random Forest is effective in obtaining the lowest maximum drawdown and a higher cumulative return than the other methods. It is able to reach the best risk-adjusted return among all methods.

CONCLUSION

This chapter establishes a link connecting data information processing and prescriptions for optimal portfolio decisions. In the era of "big data," one should take advantage of rich data resources for making effective decisions. For example, online activities such as social media about customers' reactions to launches of new products, economic data, and the up-stream or down-stream supplier's and retailer's activities, all of which provide insights on the profitability of a company, and thus are directly or indirectly associated with the stock price of a company.

We demonstrate that machine learning methods help to improve the out-of-sample performance of data-driven portfolio decisions with drawdown constraints. Our simulation study shows that out of sample performances of portfolios utilizing machine learning outperform the benchmark portfolio. In particular, among the three implemented machine learning methods, the Random Forest consistently outperforms other methods in all measures of performance metrics. We also notice that as the training size increases, the benefits of applying machine learning methods for data-driven portfolio decisions manifest in the improvements of out-of-sample performances. On the contrary, portfolios constructed purely based on a sample path of past returns have poorer out-of-sample performance than those utilizing the machine learning methods.

In summary, our work shows a positive outlook for practitioners who are interested in applying machine learning methods to process data information and derive data-driven optimal investment decisions.

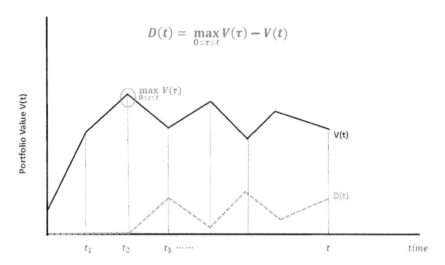

Figure 5.1. Illustration of Drawdown: the black curve represents the time series of portfolio value. The dashed curve represents the time series of drawdown. The drawdown measures the decline from the peak to the trough.

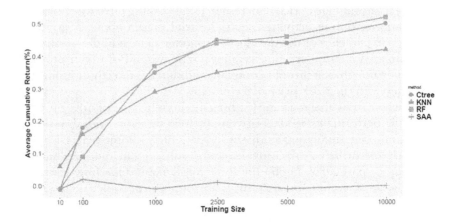

Figure 5.2.   Portfolio average cumulative return on validation data. For each realization we compute the portfolio cumulative return during the validation period. We then compute the average cumulative return over the 100 realizations. We focus on portfolio with maximum drawdown constraint and constructed based on six sizes of train data: 10, 100, 1000, 2500, 5000, 10000.

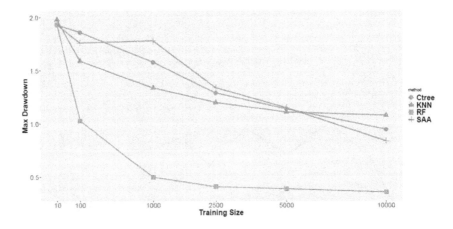

Figure 5.3.   Portfolio maximum drawdown on validation data. For each realization we compute the maximum drawdown of portfolio returns during the validation period. We then compute the average maximum drawdown over the 100 realizations. We focus on portfolio with maximum drawdown constraint and constructed based on six sizes of train data: 10, 100, 1000, 2500, 5000, 10000.

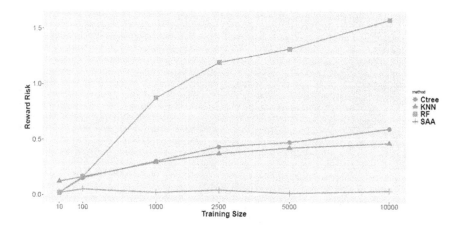

Figure 5.4. Portfolio reward risk on validation data. For each realization we compute the ratio of the cumulative return to the maximum drawdown of portfolio returns during the validation period. We then compute the average reward-risk over the 100 realizations. We focus on portfolio with maximum drawdown constraint and constructed based on six sizes of train data: 10, 100, 1000, 2500, 5000, 10000.

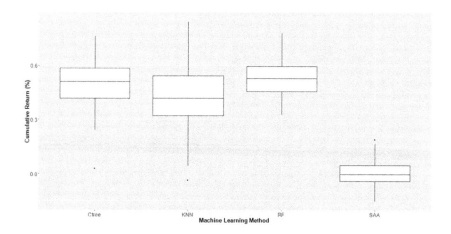

Figure 5.5. Distribution of portfolio cumulative returns on validation data based on different machine learning methods. We focus on portfolio with maximum drawdown constraint and constructed based on the size of 10000 in the training data.

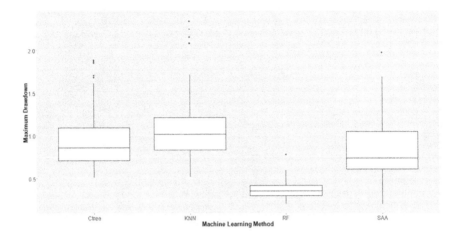

Figure 5.6.   Distribution of portfolio maximum drawdown during the validation period based on different machine learning methods. We focus on portfolio with maximum drawdown constraint and constructed based on the size of 10000 in the training data.

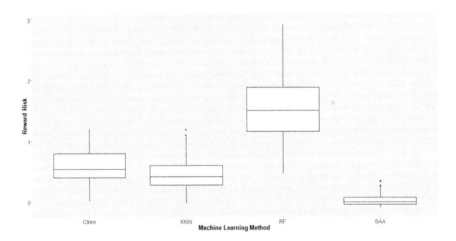

Figure 5.7.   Distribution of portfolio risk reward ratio during the validation period based on different machine learning methods. We focus on portfolio with maximum drawdown constraint and constructed based on the size of 10000 in the training data.

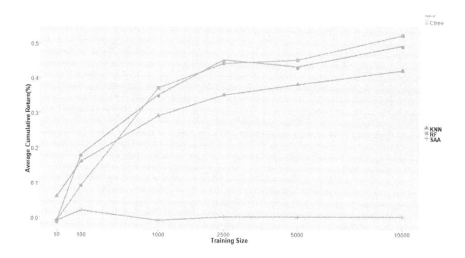

Figure 5.8.   Portfolio average cumulative return on validation data. For each realization we compute the portfolio cumulative return during the validation period. We then compute the average cumulative return over the 100 realizations. We focus on portfolio with $CDaR_{\alpha=0.05}$ constraint and constructed based on six sizes of train data: 10, 100, 1000, 2500, 5000, 10000 and a fixed size of 1000 in the validation data.

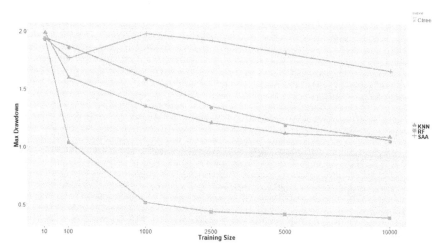

Figure 5.9.   Portfolio maximum drawdown on validation data. For each realization we compute the maximum drawdown of portfolio returns during the validation period. We then compute the average maximum drawdown over the 100 realizations. We focus on portfolio with $CDaR_{\alpha=0.05}$ constraint and constructed based on six sizes of train data: 10, 100, 1000, 2500, 5000, 10000.

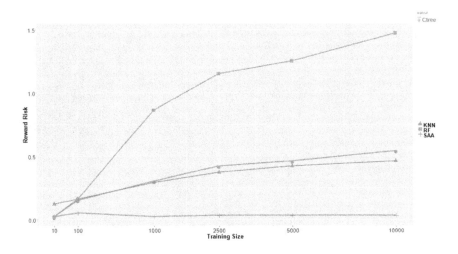

Figure 5.10. Portfolio reward risk on validation data. For each realization we compute the ratio of the cumulative return to the maximum drawdown of portfolio returns during the validation period. We then compute the average reward-risk over the 100 realizations. We focus on portfolio with with $CDaR_{\alpha=0.05}$ constraint and constructed based on six sizes of train data: 10, 100, 1000, 2500, 5000, 10000.

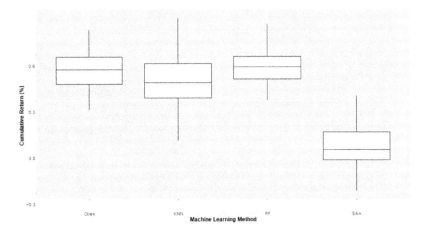

Figure 5.11. Distribution of portfolio cumulative returns during the validation period based on different machine learning methods. We focus on portfolio $CDaR_{\alpha=0.05}$ constraint and constructed based on the size of 10000 in the training data.

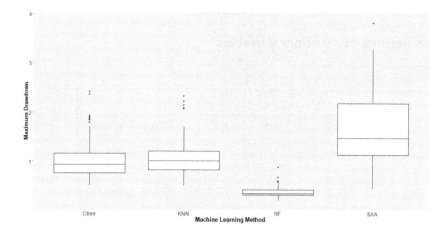

Figure 5.12. Distribution of portfolio maximum drawdown during the validation period based on different machine learning methods. We focus on portfolio with $CDaR_{\alpha=0.05}$ constraint and constructed based on the size of 10000 in the training data.

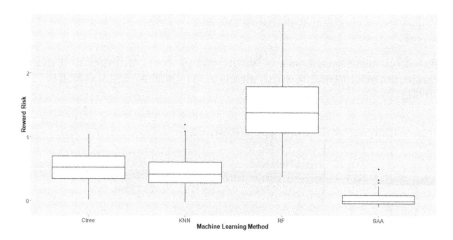

Figure 5.13. Distribution of portfolio risk reward ratio during the validation period based on different machine learning methods. We focus on portfolio with $CDaR_{\alpha=0.05}$ constraint and constructed based on the size of 10000 in the training data.

APPENDIX

Simulations of Auxiliary Variables

$$\Phi_1 = \begin{pmatrix} 0.5 & 0.9 & 0.1 \\ 1.1 & -0.7 & 0 \\ 0 & 0 & 0.5 \end{pmatrix}, \quad \Phi_2 = \begin{pmatrix} 0 & -0.5 & 0 \\ -0.5 & 0 & 0 \\ 0 & 0 & 0 \end{pmatrix}$$

$$\Theta_1 = \begin{pmatrix} -0.4 & -0.8 & 0.0 \\ 1.1 & 0.3 & 0 \\ 0 & 0 & 0.0 \end{pmatrix}, \quad \Theta_2 = \begin{pmatrix} 0 & 0.8 & 0 \\ 1.1 & 0 & 0 \\ 0 & 0 & 0 \end{pmatrix}$$

$$(\Sigma_U)_{ij} = \left(\mathbb{1}_{[i=j]} \frac{8}{7} - (-1)^{i+j} \frac{1}{7} \right) 0.05$$

Based on these values, as $t \to \infty$, the F(t) process converges to a multi-variate normal distribution with means $\mu_F = (-0026401684, 0.0003617125, -0.0031036954)$) and covariances

$$\text{cov}_F = \begin{pmatrix} 0.22073410 & 0.08601150 & -0.01527413 \\ 0.08601150 & 0.27050837 & 0.00368179 \\ -0.01527413 & 0.00368179 & 0.06772540 \end{pmatrix},$$

which allows us to simulate the multivariate time series processes F(t).

Specifications of Model Parameters

$$
A = 0.025 \times \begin{pmatrix}
0.8 & 0.1 & 0.1 \\
0.1 & 0.8 & 0.1 \\
0.1 & 0.1 & 0.8 \\
0.8 & 0.1 & 0.1 \\
0.1 & 0.8 & 0.1 \\
0.1 & 0.1 & 0.8 \\
0.8 & 0.1 & 0.1 \\
0.1 & 0.8 & 0.1 \\
0.1 & 0.1 & 0.8 \\
0.8 & 0.1 & 0.1 \\
0.1 & 0.8 & 0.1 \\
0.1 & 0.1 & 0.8
\end{pmatrix}, \quad
B = 0.075 \times \begin{pmatrix}
0 & 1 & 1 \\
1 & 0 & 1 \\
1 & 1 & 0 \\
0 & 1 & 1 \\
1 & 0 & 1 \\
1 & 1 & 0 \\
0 & 1 & 1 \\
1 & 0 & 1 \\
1 & 1 & 0 \\
0 & 1 & 1 \\
1 & 0 & 1 \\
1 & 1 & 0
\end{pmatrix}
$$

REFERENCES

Ban, G., Karoui, N., & Lim, A. (2018). Machine learning and portfolio optimization. *Management Science, 64*(3), 983–1476.

Bertsimas, D., & Kallus, J. (2020). From perdictive to prescriptive analytics. *Management Science, 66*(3), 1025–1044.

Breiman, L. (2001). Random Forest. *Machine Learning, 45*(1), 5–32.

Cvitanic, J., & Karatzas, I. (1995). On portfolio optimization under drawdown constraints. *IMA Lecture Notes in Mathematics and Application, 65*, 77–88.

Chekhlov, A., Uryasev, S., & Zabarankin, M. (2003). Portfolio optimization with drawdown constraints. In B. Scherer (Ed.), *Asset and liability management tools.* Risk Book.

Chekhlov, A., Uryasev, S., & Zabarankin, M. (2005). Drawdown measure in portfolio optimization. *International Journal of Theoretical and Applied Finance, 8*(1), 13–28.

Grossman, S. J., & Zhou, Z. (1993). Optimal investment strategies for controlling drawdowns. *Mathematical Finance, 3*, 241–276.

Hothorn, T., Hornik, K., & Zeiles, K. (2006). Unbiased recursive partitioning: A conditional inference framework. *Journal of Computational and Graphical Statistics, 153*, 651–674.

CHAPTER 6

MINING FOR FITNESS

Analytical Models That Fit You so You Can Be Fit

William Asterino and Kathleen Campbell
Saint Joseph's University

ABSTRACT

Evaluating body mass index (BMI) and weight are conventional methods of assessing overall health and fitness as well as risk issues, such as heart disease. Using recent identified and verified trends, this chapter considers adjusting the dependent measure to waist circumference as a preferred indicator of identifying change in health and fitness. This exploratory analysis considers two models: (1) a logistic regression technique initialized with 23 independent variables (reduced to 10 variables) and (2) a multiple linear regression model (reduced to 9 variables). Each mixed model can be used to predict changes in waist circumference efficiently. Both models consisted of 6 overlapping variables. A small external sample is applied to assess the validity of each model. Overall, these models can be used as a starting point to assess risk factors which would begin a trend towards unhealthy lifestyle while considering classic variables such as calories, sleep duration, exercise method, fasting, and consumption indicators (carbs, proteins, fats, caffeine, and alcohol).

Contemporary Perspectives in Data Mining, Volume 4, pp. 77–99

INTRODUCTION

Every fitness regime starts somewhere. Common possibilities include: being healthier, looking better, or a combination of the two. Often, people set out to improve their overall fitness level and look to current fads or consistent staples to get started. The health and wellness community is rich with information pertaining to a variety of diets, exercise routines, and heuristics that usually lack scientific backing. Despite this plethora of information, a prospective fitness enthusiast is often left confused as to the best direction when initiating their fitness journey. This results in a never-ending system of trial-and-error until an individual either subscribes to a method or gives up altogether.

Achieving physical fitness and wellness can be overwhelming as a variety of nutritional and exercise options are available. There are a multitude of diets with varying ratios of macronutrients, such as ketogenic diets consisting of very low consumption of carbs and paleolithic diets consisting of consuming whole foods that would have been available to humans during the paleolithic era (Freedman & King, 2001). There is also the fast-growing popularity of vegan diets (Turner-McGrievy et al., 2012) which is a diet based on consuming only whole plants, such as fruit, vegetables, legumes, nuts, and seeds. There are multiple exercise methods including resistance training, low-intensity cardio, high-intensity interval training, plyometrics, and calisthenics. Some individuals even experiment with meal timing using methods such as intermittent fasting, which is growing in popularity as a "magic pill" for body fat reduction as some research suggests (Moro et al., 2016). Each of these methods is effective in its own way, yet it is difficult to pinpoint a singular optimal method as contradictions prevail among most methods and diets.

In this case study, elementary statistics through advanced analytics (multiple linear regression and logistic regression) are applied in an attempt to address which variables are optimal for predicting changes in waist size. To do this, various data points pertaining to one prolonged set of wellness routines were tracked in an attempt to pinpoint the best exercise methods in combination with diet. A training set including 100 observations is used to create two comparable models which identified ten and nine significant variables, respectively. A training set was then applied for validation and model comparison.

The dependent variable (waist circumference) was chosen over weight and body mass index due to several studies which confirm it is a stronger indicator of body composition and overall health when compared to the others (Janssen et al., 2004; Obesity Education Initiative, n.d.; Pouliot et al., 1994). Once this dependent variable was chosen, over 30 health and wellness articles were considered to help identify the most popularly considered independent variables for collection, hence guiding the variables considered for this analysis. The common themes that appeared throughout the abstracts of these articles were considered. A quick view of common themes appears within the word-cloud (see Figure 6.1) where the increased size of words such as body fat, calories, and caffeine are due to the fact that these words are more frequently stated within the combined abstracts. Figure 6.1 displays a combination of prominent words, diets, and weight-loss regimes.

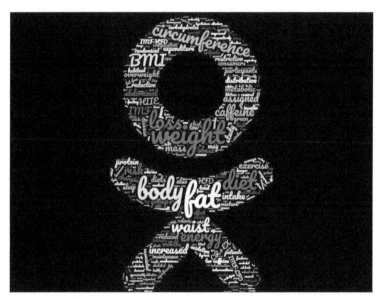

Figure 6.1.   Word cloud considering 32 article abstracts dealing with fitness.

Data

The selected study variables, such as caloric intake, macro/micronutrients (carbs, protein, fats, saturated fats, and sugar), exercise method, rest time, and the dependent variable change in waist circumference were evident throughout the articles (see Figure 6.1). Change in waist circumference was selected as the dependent variable since it is easily measurable and considered a better indicator for body fat percentage and overall fitness than other indicators (Janssen et al., 2004; Pouliot et al., 1994), such as weight or BMI. Change in waist circumference was determined by subtracting the waistline measurement of the previous day from the current day to determine the change over the course of the day. Sleep duration was selected in this study after considering its effectiveness observed in other research studies (Nedeltcheva et al., 2010; Pouliot et al., 1994). Additional variables such as caffeine consumption, daily fasting duration, water consumption, and alcohol consumption were recorded because there is little literature on how these variables affect health and fitness. Exercise method (resistance training, low-intensity cardio, and high-intensity cardio) and rest were recorded as binary variables that were valued as 1 if the activity was performed over the course of the day. The selected variables were recorded over 100 days, and the data was collected from a sample of an individual male, aged 21–22 years, following a set of fitness exercises and daily routines and tested by 52 data points, which were collected from the initial individual and 4 people across 5 weeks with varying lengths (see Table 6.1).

Table 6.1
Table Providing Pertinent Individual Statistics, Model Accuracy, and Comparison and Type of Data Sets

Person ID	Days Collected	Gender	Age	Height (inches)	LR Model Accuracy	MLR Model Accuracy	Overlap Between Models	Training/ Testing
1	100	Male	22	69	80% (80/100)	85% (85/100)	66% (66/100)	Training
1	32	Male	22	69	81% (26/32)	69% (22/32)	78% (25/32)	Testing
2	5	Male	22	73	80% (4/5)	40% (2/5)	60% (3/5)	Testing
3	5	Male	22	70	80% (4/5)	60% (3/5)	80% (4/5)	Testing
4	5	Male	22	69	80% (4/5)	80% (4/5)	60% (3/5)	Testing
5	5	Female	21	66	100% (5/5)	100% (5/5)	100% (5/5)	Testing

For the initial data collected, the distribution of each variable was considered (shape, spread, and outliers) and feasible correlations among each set of paired variables were examined to assess if any issues should be addressed prior to analysis. The normal continuous variables (see Table 6.2) include change in waist circumference, water consumption, sleep duration, and daily fasting duration. The dependent variable, change in waist circumference, contained a negative value (decrease or no change in waist circumference) over 63 days of the 100-day period of initial data collection, while change in waist circumference was positive (increase) over 37 days of the period. Change in waist circumference was normally distributed over an average of –0.04 inches, while the standard deviation was 0.26 inches, or about ¼ inch (see Figure 6.2). Outliers were identified for the normal variables using mean and standard deviation (see Table 6.2). It is important to note that there were 100 observations and approximately 5% of data fell outside of two standard deviations (95% confidence), which is expected. Therefore, to identify true outliers, three standard deviations was used for Table 6.2.

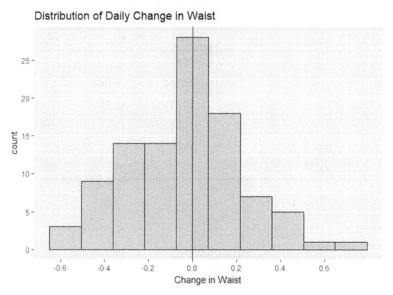

Figure 6.2. Distributions of change in waist measurement day over day for collected data.

Table 6.2
Normal Continuous Variables Relative Statistics Along With Feasible Outliers

Variable	Mean	Standard Deviation
Change in WC	-0.04	0.26
Water	9.86	2.07
Sleep	6.97	0.84
Fasting	16.34	3.44

The nutritional variables (calories, carbs, protein, fats, saturated fats, and sugar) were hill shapes with slight-right kurtosis. The kurtosis was identified as "cheat days" which involve the overconsumption of calorically-dense foods over the course of a day as a deviation from typical dietary routines. These days remain in the data as they are considered part of real-life consumption routines. The median, quartiles, and interquartile range were used to identify outliers for the skewed data (see Table 6.3). For consistency 3 IQR were considered.

Table 6.3
Variables That Revealed Skewness Along With Feasible Outliers Identified

Variable	Median	Q1	Q3	IQR	Outliers
Calories	2111	1813	2587	774	
Carbs	224	173.75	283.50	109.75	
Protein	112	84.75	127.25	42.50	
Fats	93	73.75	114.75	41	245, 310
Sat. Fats	30	24	41	17	152
Sugar	49	33	77	44	276

Table 6.4
Binary Variables

Binary Variable	Days of Occurrence (%)	Days of No Occurrence (%)
Caffeine Consumption	81	19
Alcohol Consumption	41	59
Resistance Training	26	74
Low-Intensity Cardio	25	75
High-Intensity Cardio	18	82
Rest/No Exercise	53	47

The distribution of binary variables, such as resistance training, low-intensity cardio, high-intensity cardio, rest, caffeine consumption, and alcohol consumption, were identified in terms of the proportion that they occurred over the course of the study. Exercise was performed over 47% of the 100-day study, while rest days made up 53% of the study. The breakdown of exercise method during the exercise portion was 55%, 53%, and 38% for resistance training, low-intensity cardio, and high-intensity cardio, respectively (Interaction among these variables was built in as different types of exercises overlap on predetermined days.) Caffeine of some type was consumed during 81% of the study, and alcohol was consumed during 41% of the study. The count of each binary variable is located in Table 6.3 (The distributions and measures of center and spread for all variables are located in the Appendix.)

The correlations were examined to assess any relationships among the variables. Some notable correlations associated with the dependent variable (waist circumference) are a 0.32 correlation between change in waist circumference and caloric intake (increased calories produce increased circumference), and a –0.38 correlation between change in waist circumference and sleep duration (increased sleep is associated with decreased circumference). These correlations agree with the findings cited in several research studies (Redman et al., 2007; Sacks et al., 2009; Taheri et al., 2004). The correlations between change in waist circumference and macronutrients (carbs, protein, fats, saturated fat) were positively correlated, primarily because these are components of caloric intake. Interestingly, carbs were more strongly correlated with an increase in waist circumference (correlation = .30); perhaps because carbs encompass sugar intake as well (Malik et al., 2006). Other significant correlations include a moderately negative correlation between caloric consumption and daily fasting duration and strong positive correlations between caloric consumption and macronutrients, which confirms a known relationship (Sacks et al., 2009). A detailed correlation matrix considering all the variables is located in the Appendix.

Following rigorous examination, these variables were deemed robust enough for predictive modeling to determine a method that could accurately predict change in waist circumference.

Multiple Linear Regression Versus Median Logistic Regression

Two models were developed to help predict change in waist circumference: a multiple linear regression and a median logistic regression. In analyzing the data, a primary consideration was to identify whether similar variables would arise in both analyses. An overall model comparison was made to see how well each model predicted the change in waist circumference and compare model accuracy. Model accuracy was defined as identifying how often each model correctly predicted if waistline increased or decreased/ no change daily. The purpose of a multiple linear regression is to model the relationship between one or more explanatory variables and a continuous response/dependent variable. This type of model generates a linear equation that is tailored to the relationship between the observed explanatory variables and the dependent variable. The general formula for a multiple linear regression is summarized below in which $\hat{y}$ is the dependent variable, β_0 is the y-intercept of the equation, while β_n is the slope coefficient or the effect on for each additional unit of the explanatory variable x_n added as input when all else is held constant:

$$\hat{y} = \beta_0 + \beta_1 x_1 + \beta_2 x_2 + \dots \beta_n x_n + E$$

In this case, the $\hat{y}$ was defined as the change of waist circumference for the current day minus the previous day with the variables considered being the previous day's exercise, nutritional, and sleep regiment. For example, if the $\hat{y}$ for a given day is 0.25, this implies that the model is predicting waist circumference to increase 0.25 inches between the current and previous day's measurements given the exercise, nutritional, and sleep regiment variables (measured the previous day) as inputs. Likewise, if the $\hat{y}$ for a given day is –0.25, this implies that the model is predicting waist circumference to decrease 0.25 inches between the current and previous day's measurements given the exercise, nutritional, and sleep regiment variables as inputs After examining the model output, a new variable was created for comparison purposes where the continuous $\hat{y}$ variable is recoded as a binary variable. If the $\hat{y}$ denoted an increase in waist circumference, it was given the value 1. Conversely, if the $\hat{y}$ denoted a decrease or no change in waist circumference, it was given the value 0. This additional variable enables the multiple linear regression and logistic regression models to be easily comparable in terms of directional accuracy since a logistic regression is binary by nature. When considering the output for model comparison, any time that the predicted value identified an increase in waist circumference and the actual change in waist circumference was an increase OR the predicted was a decrease and the actual was a decrease was considered as a success (1) versus opposing directions between predicted and actual. This allows for a direct comparison for each response when comparing the multiple linear regression to the logistic regression for accuracy. Overall, this allows for the assertion of consistency in prediction when comparing the models.

Similar to a multiple linear regression, logistic regression identifies the relationship between one or more explanatory variables and a response variable. However, a logistic regression differs from a multiple linear regression by predicting the response variable as a binary value. It does this by generating the response variable as a probability between 0 and 1, unlike a multiple linear regression which may generate a continuous number. When modelling and interpreting logistic regression, a cutoff point is selected among the range of probabilities. Any predicted probability that is above the cutoff point is given the value 1. Conversely, any predicted probability below the cutoff point is given the value 0. This cutoff point is optimized to most accurately predict whether a response will be a 0 or 1. When the median among the range of probabilities is selected as the cutoff point, it becomes a median logistic regression. The equation for a logistic regression is summarized below in which $\hat{p}$ is the response variable as a probability, $\beta_0 + \beta_1 x_1 + \beta_2 x_2 + \ldots \beta_n x_n$ is the linear equation developed by the model to fit the observed data to the response variable, and the exponential transformation to probability can be seen with the logit function:

$$\hat{p} = \frac{e^{\beta_0 + \beta_1 x_1 + \beta_2 x_2 + \ldots \beta_n x_n}}{1 + e^{\beta_0 + \beta_1 x_1 + \beta_2 x_2 + \ldots \beta_n x_n}}$$

Interpreting the accuracy of the model output was accomplished for each model by comparing sensitivity and specificity. Sensitivity is essentially defined as the number of true positives that a model is able to generate or diagnose. Sensitivity was used when interpreting the logistic regression to determine any successes when predicting increases in waist circumference. For instance, any time the logistic regression predicted an increase in waist circumference and the actual change in waist circumference was an increase was considered a success. Unlike sensitivity, specificity is used to identify the number of true negatives that a model is able to generate or diagnose. Specificity was also used when interpreting the logistic regression to determine successes associated with decreases/no change in waist circumference. As a result, any time the logistic regression predicted a decrease in waist circumference and the actual change in waist circumference was a decrease was considered a success. Conversely, any time the predicted value and actual value did not match directionally (predicted was an increase and actual was a decrease or predicted was a decrease and actual was an increase) was considered a failure similar to the multiple linear regression.

Multiple Linear Regression Model

In terms of independent variables, the final multiple linear regression retained several continuous variables, including caloric intake, consumption of protein, fats, and saturated fats, as well as daily fasting duration. The model included categorical variables that were transformed into dummy variables to conduct the analysis. These variables include alcohol (yes =1, no = 0) consumption, exercise type (resistance training, low-intensity cardio, or high-intensity cardio), and sleep duration less than 7 hours on a given day. Although sleep duration was not present as a continuous variable in this model, there was a significant categorical variable determining whether sleep duration was less than 7 hours (1 indicates less than 7 hours of sleep; 0 indicates more than 7 hours of sleep). The dependent variable for the model was change in waist circumference in inches as a continuous variable. This variable was created by subtracting the waistline measurement from the previous day from that of the current day to determine change over the course of the day. The formula developed from the model and the regression output are summarized below:

Multiple Linear Regression Equation

Change in Waist Circumference (inches) = –0.6324 + 0.00021(Cal) – 0.00094(Protein) – 0.00144(Fats) + 0.00316(SFA) + 0.00953(Fasting) – 0.10090(Alcohol) + 0.11030(Resistance Training) – 0.02801(Low-Int Cardio) + 0.12820(High-Int Cardio) + 0.23490(Less than 7 Hours)

Table 6.5
Multiple Linear Regression Output

Variable	Coefficient	P-Value	Significant in Logistic Regression?
Calories	0.00021	0.0000	Yes
Protein	–0.00094	0.2207	Yes
Fats	–0.00144	0.0191	Yes
Sat. Fats	0.00316	0.0263	Yes
Fasting Duration	0.00953	0.1603	Yes
Alcohol	–0.10090	0.0393	Yes
Resistance Training	0.11030	0.0780	No
Low-Intensity Cardio	–0.02801	0.6563	No
High-Intensity Cardio	0.12820	0.0257	No
Sleep: Less than 7 Hours	0.23490	0.0000	No

When examining the regression output, it is apparent that the multiple linear regression shares several variables with the logistic regression, such as calories, protein, fats, saturated fats, fasting duration, and alcohol, and also includes exercise method and sleep as a categorical variable (not present in the logistic regression). The coefficient for calories 0.021 inch increase for each additional 100 calories consumed or a 0.21 inch increase for every additional 1,000 calories consumed. While calories are accounted for in the model, the protein, fat, and saturated fat variables represent the impact that the macronutrient ratio has on waist circumference, such as the potential impact of eating more protein and less fat versus less protein and more fat. The coefficients for protein, fat, and saturated fat affect circumference –0.00094, –0.00144, and 0.00316, respectively for each additional gram of protein, fat, and saturated fat. More realistically, it can be interpreted that the impact on waist circumference for every additional 100 grams consumed for each variable. For every additional 100 grams of protein consumed, waist circumference is expected to decrease 0.094 inches. Waist circumference is expected to decrease 0.144 inches for every 100 grams of fat consumed. Conversely, waist circumference is expected to increase 0.316 inches for every 100 grams of saturated fat consumed. Fasting duration (measured in hours) has a coefficient of 0.00953, which translates to a 0.00953 inch increase in waist circumference for each additional hour fasted before consumption of the first meal of the day. The alcohol coefficient of –0.1009 is interpreted that waist circumference is expected to decrease 0.1009 inches simply when alcohol is consumed during the day. The coefficients by exercise method are 0.1103, –0.02801, and 0.1282 for resistance training, low-intensity cardio, and high-intensity cardio, respectively. These represent the impact that performing each exercise method has on waist circumference. When resistance training is performed, waist circumference is expected to increase 0.1103 inches, while waist circumference is expected to decrease 0.02801 inches when low-intensity cardio is performed. Waist circumference is expected to increase 0.1282 inches when high-intensity cardio is performed. The coefficient for sleep: less than 7 hours is 0.2349, which translates to an expected 0.2349 inch increase in waist circumference when sleep duration is less than 7 hours. The model's sensitivity (true positives) of 96% and specificity (true negatives) of 70% result in an overall accuracy of 85% for the training data.

Median Logistic Regression Model

In terms of independent variables, the final logistic regression retained several continuous variables, including caloric intake, consumption of carbs, protein, fats, and saturated fats, as well as daily fasting duration and sleep duration (carbs and sleep duration were not significant in the multiple linear regression) (see Table 6.3). The model included categorical variables that were transformed into dummy variables to conduct the analysis. These variables (yes = 1, no = 0) include caffeine and alcohol consumption. The dependent variable for the model was change in waist circumference as a binary variable in which "1" denotes an increase in waist circumference and "0" denotes a decrease or no change in waist circumference. The formula developed from this model generates the probability that waist circumference will decrease or not change ($\hat{p}$), and is outlined below:

$$p = \frac{e^{4.98-0.005516(Cal)-0.020(Carbs)-0.020(Protein)-0.048(Fats)-0.079(Sat Fats)-1.70(Caffeine)-1.69(Sleep)-0.27(Fasting)-2.77(Alcohol)}}{1-e^{4.98-0.005516(Caloric)-0.020(Carbs)-0.020(Protein)-0.048(Fats)-0.079(Sat Fats)-1.70(Caffeine)-1.69(Sleep)+0.27(Fasting)-2.77(Alcohol)}}$$

The equation above generates a probability between 0 and 1. When examining the distribution of probabilities from the experiment, the cutoff point for determining whether waist circumference would increase (1) or decrease/stay the same (0) was 0.316, the median of the probability distribution. This produced the optimal cutoff point for accurately predicting whether waist circumference would decrease, specifically. Below is the distribution of probabilities generated by the model as well as the cutoff point of 0.316 outlined by the vertical line. The graphic shows that the distribution of probabilities to the left of the vertical line were denoted by the model as decreases/no change (0) in waist circumference (higher proportion of pink is evidence of accuracy), while the distribution of probabilities to the right of the line were denoted as increases in waist circumference (1) (higher proportion of blue is evidence of accuracy). The graphic visually reveals the model sensitivity (true positives) of 92% and specificity (true negatives) of 68% for the training data.

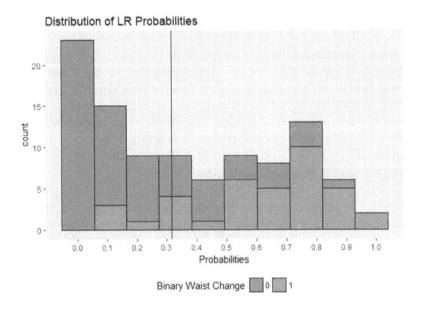

Figure 6.3

Table 6.6
Median Logistic Regression Output (for Waist Change Odds of 0 Versus 1)

Variable	Unit Odds Ratio	Prob > ChiSq	Significant in Multiple Linear Regression?
Calories	0.9944	0.0009	Yes
Carbs	1.0202	0.0096	No
Protein	1.0203	0.0777	Yes
Fats	1.0495	0.0077	Yes
Sat. Fats	0.9237	0.0003	Yes
Caffeine	5.463	0.0274	No
Sleep Duration	5.394	0.0000	No
Fasting Duration	0.7604	0.0041	Yes
Alcohol	15.99	0.0012	Yes

The logistic regression output in Table 6.3 shows the odds that waist circumference will decrease or not change for each additional unit added of the independent variables. The unit odds ratios are interpreted that any unit odds ratio above 1 is associated with an increase in odds that waist circumference will decrease/not change, while any unit odds ratio below 1 is associated with a decrease in odds that waist circumference will decrease/not change. The difference between the unit odds ratio and 1 is the percent increase/decrease in odds for each additional unit added. For instance, calories has a unit odds ratio of 0.9944. This means for each additional calorie consumed, the odds that waist circumference will decrease/ not change decrease by 0.0056 or 0.56%. The unit odds ratios for carbs, protein, and fats are 1.0202, 1.0203, and 1.0495, respectively. This may be interpreted that the odds waist circumference decreasing/not changing increase 2.02% for each additional gram of carb consumed, increase 2.03% for each additional gram of protein consumed, and increase 4.95% for each additional gram of fat consumed. The unit odds ratio of 0.9237 for saturated fats translates to 7.63% decrease in odds that waist circumference will decrease/not change for each additional gram of saturated fat consumed. In terms of fasting duration, the odds of waist circumference decreasing/not changing decrease 23.96% for each additional hour fasted before consuming first meal of the day. Since caffeine has a unit odds ratio of 5.463, the odds that waist circumference will decrease/not change increase 5.463 times on days when any type of caffeine is consumed. Likewise, sleep duration has a unit odds ratio of 5.394, meaning the odds that waist circumference will decrease/not change increase 5.394 times for each additional hour slept. The unit odds ratio of 15.99 for alcohol translates to increased odds that waist circumference will decrease/not change by 15.99 times when any type of alcohol is consumed during the day.

RESULTS AND MODEL COMPARISON

When examining the results of the multiple linear regression model, caloric intake is associated with an increase in waist circumference and an overall reduced caloric intake leads to a higher likelihood of a decrease in waist circumference. Although calories are accounted for in the model, the protein, fat, and saturated fat variables encompass the impact that macronutrient ratios have on the change in waist circumference, such as the impact of consuming more protein over more fat. The model shows that there is no apparent benefit to consuming more protein over more fat, or vice versa. However, consuming saturated fats is associated with an increase in waist circumference so it may be preferable to consume more unsaturated fats over saturated fats (Piers et al., 2003). Daily fasting duration is positively correlated with an increase waist circumference. Consumption of alcohol, regardless of the type, is associated with a decrease in waist circumference. In terms of exercise method, performing resistance training or high-intensity cardio are associated with an increase in waist circumference, possibly associated with the temporary inflammation that is present after exercise. Although sleep duration was not present as a continuous variable in the model, sleep duration less than 7 hours as a categorical variable was associated with an increase in waist circumference. The model was tested on additional data and the residuals were calculated. When examining the testing data, the sensitivity (true positives) and specificity (true negatives) were identified as 82.8% and 52.2%, respectively. The distribution of the residuals was normal with the majority of values concentrated around 0. The mean of the residuals was approximately 0 with a standard deviation of 0.195. As a result, this method may be considered an adequate model. The image below shows the distribution of residuals for the multiple linear regression. Generally, when the predicted change in waist circumference was less than the actual change in waist circumference, waist circumference decreased or did not change in most cases (higher proportion of pink to the left of zero indicates accuracy of model). Conversely, when the predicted change was greater than the actual change, it was more likely that waist circumference increased (higher proportion of blue to the right of zero indicates accuracy of model).

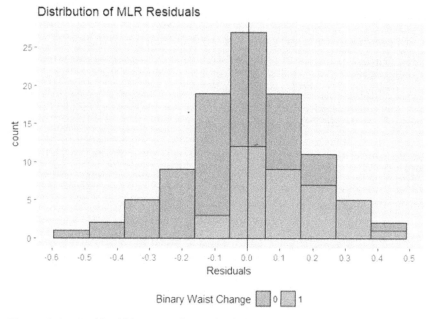

Figure 6.4.   Residual histogram for multiple linear regression model.

When examining the results of the logistic regression model, the sensitivity (true positives) and specificity (true negatives) for the testing data were taken into account. The sensitivity of the model was 86.1%, meaning that the model correctly predicted if waist circumference would decrease/ not change 86.1% of the time. The specificity of the model was 75%, meaning that the model correctly predicted whether waist circumference would increase 75% of the time. Overall, this model correctly predicted if waist circumference would increase or decrease/not change 82.7% of all instances. In terms of takeaways from the model, it produced similar results to the multiple linear regression model. Caloric intake was positively correlated with increased odds of waist circumference increasing. There was no apparent difference in the ratio of macronutrients consumed since the weightings of carbs, protein, and fats were close to equal in this model. However, increased consumption of saturated fats was associated with increased odds of waist circumference increasing. Caffeine and alcohol consumption were both associated increased odds of waist circumference decrease/no change. Daily fasting duration was associated with increased odds of waist circumference increase. Sleep duration, the most significant variable in this model, was negatively correlated with waist circumference and is associated with increased odds that waist circumference would decrease/not change. This essentially means that longer sleep duration translates to a greater likelihood that waist circumference will decrease/not change.

Overall these two models closely align with each other. Increased caloric intake is associated with an increase in waist circumference, while reduced caloric intake is associated with a decrease/no change in waist circumference. Sleep duration and consumption of alcohol are associated where an increase of each provides a decrease/no change in waist circumference in both models. Fasting duration is associated with an increase in waist circumference in both models. Both models also retain consumption of protein, fat, and saturated fat as significant variables. Essentially, the models only differ in that the multiple linear regression retains exercise type as significant variables, while the logistic regression retains carbs and caffeine consumption as significant variables.

When examining the results of the models (see Table 6.4), it was determined that there was a fair amount of overlap in the predictions between the two models. Within the testing data, both models were consistent and generated the same prediction in 77% of all observations (Person 1 – 78%, Person 2 – 60%, Person 3 – 80%, Person 4 – 60%, Person 5 – 100%). Overall, the logistic regression correctly predicted whether waist circumference would increase or decrease/not change 82.7% of all observations in the testing set versus the multiple linear regression that made correct predictions 69.2% of all observations in the testing set. The sensitivity and specificity for both models were also considered as metrics for comparison. The logistic regression produced a sensitivity of 86.1% while the multiple linear regression produced a sensitivity of 82.8%. Likewise, the specificity of 75% for the logistic regression was higher than the specificity of 52.2% for the multiple linear regression. These results reveal that the logistic regression was more accurate than the multiple linear regression in predicting whether waist circumference would increase or decrease/not change. Although both models produced strong results, the logistic regression appears to outperform the multiple linear regression for all individuals within the testing data. As a result, the logistic regression may be considered a preferable technique when predicting change in waist circumference.

CONCLUSIONS AND FINDINGS

This analysis resulted in a variety of findings, namely the importance of caloric intake and sufficient sleep duration, as well as the insignificance of macronutrient ratios and exercise type in reducing waist circumference. It must be noted that this is not an argument for one ideal method or formula

Table 6.4
Comparison Between Multiple Linear Regression and Logistic Regression

Person ID	Days Collected	LR Model Accuracy	MLR Model Accuracy	Overlap Between Models	Training/ Testing
1	100	80%	85%	66%	Training
1	32	81%	69%	78%	Testing
2	5	80%	40%	60%	Testing
3	5	80%	60%	80%	Testing
4	5	80%	80%	60%	Testing
5	5	100%	100%	100%	Testing

for a guaranteed decrease in waist circumference or weight loss. However, it does challenge recent fads, such as the superiority of high-carbohydrate or ketogenic diets over other dietary methods, disagreeing with some research touting their efficacy (Samaha et al., 2003; Shai et al., 2008). The study also supplies evidence that waistline can be linked and predicted accurately, while supporting it as an effective measure of health and fitness. Essentially, the study proves that changes in waist circumference are dependent upon a combination of decisions related to several nutritional and lifestyle indicators such as sleep duration, daily fasting duration, and whether caffeine or alcohol were consumed over the course of a day.

Although the human body is extremely complex, interesting trends and relationships were identified in the study that would allow one to have some control over their waistline measurement. At this time, a formulaic panacea is non-existent. However, the results imply that further research may help identify more robust models to help those interested in having control over this health variable. The study revealed that there are certainly indicators that are more significant than others and abiding by them is essential to realizing a change in waist circumference.

After running several statistical tests, it was concluded that some variables originally predicted to significantly influence waist circumference were indeed significant in doing so. These variables were nutritional categories such as daily caloric intake, consumption of carbs, protein, fats, and saturated fats, as well as caffeine consumption and daily fasting duration. Restricted caloric intake is generally accepted as an effective means to losing weight and changing body composition (Cox et al., 2003; Schröder et al., 2013; Stiegler & Cunliffe, 2006). In terms of macro/micronutrients, consumption of carbs, protein, fats, and saturated fats were deemed significant variables in the experiment. However, the three main macronutrients (carbs, protein, fats) were equally weighted in their effect on waist circumference, which disagrees with research endorsing consumption of specific macronutrients over others (Aller et al., 2014; Volek et al., 2004). Although varying levels of dietary fat consumption were found to be insignificant in influencing waist circumference and body fat levels (Willett & Leibel, 2002), saturated fat consumption had an adverse impact on the probability of reduction of waist circumference, which aligns with the findings of several studies (Piers et al., 2003; Wood et al., 1991). Caffeine consumption was also deemed significant in reducing waist circumference; although it is unclear as to whether there is a causal relationship between caffeine consumption and change in waist circumference or if caffeine consumption is correlated with restricted caloric intake (Westerterp-Plantenga et al., 2012). This may require further research, considering there was very little correlation between caloric intake and caffeine consumption in this study. Daily fasting duration was found to significantly influence waist circumference when examined in combination with the other variables. Although intermittent fasting is endorsed as an effective tool to become fitter and healthier, the study found that daily fasting duration had an adverse impact on waist circumference with longer fasting durations associated with a higher probability of an increase in waist circumference. These findings disagree with recent research endorsing intermittent fasting as an effective means to weight loss (Klempel et al., 2012; Moro et al., 2016). It is unclear whether there is a causal relationship between daily fasting duration and waist circumference or if daily fasting duration accounts for some variable not measured in the study, so future research should be conducted to explore the topic.

The study revealed some variables significantly influenced waist circumference to an extent that was unexpected. These variables include sleep duration and alcohol consumption. Interestingly, sleep duration had the most significant influence over waist circumference in this study, and longer sleep duration was associated with a higher probability of reducing waist circumference. These findings were consistent despite situations such as unrestricted caloric intake. These findings align with established research that claims sleep deprivation adversely impacts the metabolism and triggers hormonal changes that may lead to weight gain (Everson & Wehr 1993; Spiegel et al., 1999). The study revealed that alcohol consumption led to an increased probability of reduced waist circumference regardless of the type and amount of alcohol consumed. It is possible that alcohol consumption may lead to dehydration that artificially shrinks waist circumference, but there is very little literature on the subject and should be explored in future studies.

The study revealed variables that were expected to significantly influence waist circumference, but unexpectedly did not do so. These variables include sugar consumption, water consumption, and workout method. As an explanation, it must be noted that sugar consumption most likely appeared as an insignificant variable because its impact may be accounted for with the caloric intake and carb consumption variables (Malik et al., 2006). The finding that exercise method did not significantly influence waist circumference was unexpected in the study. Many individuals subscribe to a specific exercise method and claim for it to be the most effective way to become healthier and fitter. However, the study found that the exercise method and length in days of rest between exercise sessions did not impact the probability of waistline circumference increasing or decreasing. These findings disagree with research that prefers one exercise method over the other as a tool for weight loss (Ismail et al., 2012). However, there is additional research that support this finding, claiming exercise alone and not one particular method is effective in pursuing weight loss (Chambliss, 2005; Slentz et al., 2004). It must be noted that is important to consider other fitness goals in addition to reducing waist circumference when subscribing to an exercise method since some methods improve lean body mass while others enhance cardiovascular endurance (Willis et al., 2012).

Using the experimental data from the study, two predictive models were developed to determine whether waist circumference would increase or decrease daily. The algorithms behind the models rely upon inputs for nutritional variables such as caloric intake, consumption of carbs, protein, fats, and saturated fats, and other variables including exercise method, sleep duration, daily fasting duration, caffeine consumption, and alcohol consumption. Since the area of interest was determining the factors that reduce waist circumference, these models were tailored to properly assess if waist circumference would decrease or not change. Overall, the logistic regression may be considered the preferable technique between the two models when predicting change in waist circumference since it correctly predicted whether waist circumference would increase or decrease 82.7% of all instances. Although these models are a starting point in assessing risk factors associated with unhealthy lifestyles, this study and future studies can pave the way for the possibility of developing ideal fitness programs tailored to each individual.

REFERENCES

Aller, E., Larsen, T. M., Claus, H., Lindroos, A. K., Kafatos, A., Pfeiffer, A., Martinez, J. A., Handjieva-Darlenska, T., Kunesova, M., Stender, S., Saris, W. H., Astrup, A., & van Baak, M. A. (2014). Weight loss maintenance in overweight subjects on ad libitum diets with high or low protein content and glycemic index: the DIOGENES trial 12-month results. *International Journal of Obesity, 38,* 1511–1517.

Chambliss, H. (2005). Exercise duration and intensity in a weight-loss program. *Clinical Journal of Sport Medicine*, 15(2), 113–115. https://doi.org/10.1097/01.jsm.0000151867.60437.5d

Cox, K. L., Burke, V., Morton, A. R., Beilin, L. J., & Puddey, I. B. (2003). The independent and combined effects of 16 weeks of vigorous exercise and energy restriction on body mass and composition in free-living overweight men—A randomized controlled trial, *Metabolism: Clinical and Experimental*, 52(1), 107–115, https://doi.org/10.1053/meta.2003.50017

Everson, C. A., & Wehr, T. A. (1993). Nutritional and metabolic adaptations to prolonged sleep deprivation in the rat, *American Journal of Physiology: Regulatory, Integrative, and Comparative Physiology, 264*(2), 376–387. https://doi.org/10.1152/ajpregu.1993.264.2.R376

Freedman, M., King, J., & Kennedy, E. (2001). Popular diets: A scientific review. *Obesity Research, 9, 1S–40S.*

Ismail, I., Keating, S. E., Baker, M. K., & Johnson, N. A. (2012). A systematic review and meta-analysis of the effect of aerobic vs. resistance exercise training on visceral fat. *Obesity Reviews, 13*(1), 68–91. https://doi.org/10.1111/j.1467-789X.2011.00931.x

Janssen, I., Katzmarzyk, P. T., & Ross, R. (2004). Waist circumference and not body mass index explains obesity-related health risk. *The American Journal of Clinical Nutrition, 79*(3), 379–384. https://doi.org/10.1093/ajcn/79.3.379

Klempel, M. C., Kroeger, C. M., Bhutani, S., Trepanowski, J. F., & Varady, K. A. (2012). Intermittent fasting combined with calorie restriction is effective for weight loss and cardio-protection in obese women. *Nutrition Journal, 98.* https://doi.org/10.1186/1475-2891-11-98

Malik, V. S., Schulze, M. B., & Hu, F. B. (2006, August). Intake of sugar-sweetened beverages and weight gain: a systematic review. *The American Journal of Clinical Nutrition, 84*(2), 274–288. https://doi.org/10.1093/ajcn/84.2.274

Moro, T., Tinsley, G., Bianco, A., Marcolin, G., Pacelli, Q. F., Battaglia, G., Palma, A., Gentil, P., Neri, M., & Paoli, A. (2016). Effects of eight weeks of time-restricted feeding (16/8) on basal metabolism, maximal strength, body composition, inflammation, and cardiovascular risk factors in resistance-trained males. *Journal of Translational Medicine 2016.* https://doi.org/10.1186/s12967-016-1044-0

Nedeltcheva, A. V., Kilkus, J. M., Imperial, J., Schoeller, D. A., & Penev, P. D. (2010). Insufficient sleep undermines dietary efforts to reduce adiposity. *Annals of Internal Medicine, 153,* 435–441. https://doi.org/10.7326/0003-4819-153-7-201010050-00006

Obesity Education Initiative Electronic Textbook—Treatment Guidelines. (n.d.). *National Heart Lung and Blood Institute, U.S. Department of Health and Human Services.* www.nhlbi.nih.gov/health-pro/guidelines/current/obesity-guidelines/e_textbook/txgd/4142.htm

Piers, L. S., Walker, K. Z., Stoney, R. M. Stoney, Soares, M. J., & O'Dea, K. (2003). Substitution of saturated with monounsaturated fat in a 4-week diet affects body weight and composition of overweight and obese men. *British Journal of Nutrition, 90*(3), 717–727. https://doi.org/10.1079/BJN2003948

Pouliot, M., Després, J., Lemieux, S., Moorjani, S., Bouchard, C., Tremblay, A., Nadeau, A., & Lupien, P. J. (1994). Waist circumference and abdominal sagittal diameter: Best simple anthropometric indexes of abdominal visceral adipose tissue accumulation and related cardiovascular risk in men and women. *The American Journal of Cardiology, 73*(7), 460–468. https://doi.org/10.1016/0002-9149(94)90676-9

Redman, L. M., Heilbronn, L. K., Martin, C. K., Alfonso, A., Smith, S. R., & Ravussin, E. (2007), Effect of calorie restriction with or without exercise on body composition and fat distribution. *The Journal of Clinical Endocrinology & Metabolism, 92*(3), 865–872. https://doi.org/10.1210/jc.2006-2184

Sacks, F. M., Bray, G. A., Carey, V. J., Smith, S. R., Ryan, D. H., Anton, S. D., McManus, K., Champagne, C. M., Bishop, L. M., Laranjo, N., Leboff, M. S., Rood, J. C., de Jonge, L., Greenway, F. L., Loria, C. M., Obarzanek, E., & Williamson, D. A (2009). Comparison of weight-loss diets with different compositions of fat, protein, and carbohydrates. *New England Journal of Medicine, 360*(9), 859. https://www.nejm.org/doi/full/10.1056/NEJMoa0804748

Samaha, F. F., Iqbal, N., Seshadri, P., Chicano, K. L., Daily, D. A., McGrory, J., Williams, T., Williams, M., Gracely, E. J., & Stern, L. (2003). A low-carbohydrate as compared with a low-fat diet in severe obesity. *New England Journal of Medicine, 348*(24), 2074. https://doi.org/10.1056/NEJMoa022637i

Schröder, H., Mendez, M. A., Gomez, S. F., Fíto, M., Ribas, L., Aranceta, J., & Serra-Majem, L. (2013). Energy density, diet quality, and central body fat in a nationwide survey of young Spaniards. *Nutrition, 29*, 11–12, 1350–1355. https://doi.org/10.1016/j.nut.2013.05.019

Shai, I., Schwarzfuchs, D., Henkin, Y., Shahar, D. R., Witkow, S., Greenberg, I., Golan, R., Fraser, D., Bolotin, A., Vardi, H., Tangi-Rozental, O., Zuk-Ramot, R., Sarusi, B., Brickner, D., Schwartz, Z., Sheiner, E., Marko, R., Katorza, E., Thiery, J., Fiedler, G. M., ... Stampfer, M. J. (2008). Weight loss with a low-carbohydrate, mediterranean, or low-fat diet. *New England Journal of Medicine, 359*(3), 229–241. https://doi.org/10.1056/NEJMoa0708681.

Slentz, C. A., Duscha, B. D., Johnson, J. L., Ketchum, K., Aiken, L. B., Samsa, G. P., Houmard, J. A., Bales, C. W., & Kraus, W. E. (2004). Effects of the amount of exercise on body weight, body composition, and measures of central obesity: STRRIDE—a randomized controlled study. *Archives of Internal Medicine, 164*(1), 31–39. https://doi.org/10.1001/archinte.164.1.31

Spiegel, K., Leproult, R., & Cauter, E. V. (1999). Impact of sleep debt on metabolic and endocrine function. *The Lancet, 354*(9188), 1435–1439. https://doi.org/10.1016/S0140-6736(99)01376-8.

Stiegler, P., & Cunliffe, A. (2006). the role of diet and exercise for the maintenance of fat-free mass and resting metabolic rate during weight loss. *Sports Medicine, 36*, 239. https://doi.org/10.2165/00007256-200636030-00005

Taheri, S., Lin, L., Austin, D., Young, T., & Mignot, E. (2004). Short sleep duration is associated with reduced leptin, elevated ghrelin, and increased body mass index. *PLOS Medicine, 1*(3), e62. https://doi.org/10.1371/journal.pmed.0010062

Turner-McGrievy, G. M., Barnard, N. D., & Scialli, A. R. (2012). A two-year randomized weight loss trial comparing a vegan diet to a more moderate low-fat diet. *Obesity: A Research Journal, 15*(9), 2276–2281. https://doi.org/10.1038/oby.2007.270

Volek, J. S., Sharman, M. J., Gomez, A. L., Judelson, D. A., Rubin, M. R., Watson, G., Sokmen, B., Silvestre, R., French, D. N., & Kraemer, W. J. (2004). Comparison of energy-restricted very low-carbohydrate and low-fat diets on weight loss and body composition in overweight men and women. *Nutrition & Metabolism, 1,* Article 13. https://doi.org/10.1186/1743-7075-1-13

Westerterp-Plantenga, M. S., Lejeune, M. P., & Kovacs, E. M. (2012). Body weight loss and weight maintenance in relation to habitual caffeine intake and green tea supplementation. *Obesity: A Research Journal, 13*(7), 1195–1204. https://doi.org/10.1038/oby.2005.142.

Willett, W. C., & Leibel, R. L. (2002). Dietary fat is not a major determinant of body fat. *The American Journal of Medicine, 113*(9), Supplement 2, 47–59. https://doi.org/10.1016/S0002-9343(01)00992-5

Willis, L. H., Slentz, C. A., Bateman, L. A., Shields, A. T., Piner, L. W., Bales, C. W., Houmard, J. A., & Kraus, W. (2012). Effects of aerobic and/or resistance training on body mass and fat mass in overweight or obese adults, *Journal of Applied Physiology, 113*(12), 1831–1837. https://doi.org/10.1152/japplphysiol.01370.2011

Wood, P. D., Stefanick, M. L., Williams, P. T., & Haskell, W. L. (1991). The effects on plasma lipoproteins of a prudent weight-reducing diet, with or without exercise, in overweight men and women. *New England Journal of Medicine, 325*(7), 461. https://www.nejm.org/doi/full/10.1056/NEJM199108153250703

APPENDIX

Table 1A
Variables That Revealed Normality Along With Feasible Outliers Identified

Variable	Mean	Standard Deviation	Outliers (2-SD)
Change in WC	−0.04	0.26	−0.63, −0.58, 0.62, 0.67
Water	9.86	2.07	5, 14, 14, 14, 16
Sleep	6.97	0.84	5, 5, 5, 5.25, 9
Fasting	16.34	3.44	8, 23.25, 25.5

Table 2A
Variables That Revealed Skewness Along With Feasible Outliers Identified

Variable	Median	Q1	Q3	IQR	Outliers (1.5-IQR)
Calories	2111	1813	2587	774	3800, 3860, 4241
Carbs	224	173.75	283.50	109.75	468, 483
Protein	112	84.75	127.25	42.50	212, 216, 246
Fats	93	73.75	114.75	41	187, 191, 206, 245, 310
Sat. Fats	30	24	41	17	68, 78, 90, 152

Table 3A
Correlation Matrix among Continuous Variables

	Change in Waist	Calories
Change in Waist	1.00000000	0.31691348
Calories	0.31691348	1.00000000
Carbs	0.29683318	0.75973739
Protein	0.15198703	0.58411245
Fats	0.01769466	0.53329198
SFA	0.19135888	0.51967837
Water (8 oz serving)	0.08844301	-0.16102816
Coffee	0.08674681	0.08686564
Hours Sleep Night Before	-0.38471959	-0.01191309
Fasting Hours	0.05326150	-0.33989976
Alcohol?	-0.04037425	0.40451339

SECTION III

BUSINESS APPLICATIONS OF DATA MINING

CHAPTER 7

H INDEX WEIGHTED BY EIGENFACTOR OF CITATIONS FOR JOURNAL EVALUATION

Cuihua Hu and Feng Yang
Shanghai Lixin University of Accounting and Finance

Feng Yang and Xiya Zu
University of Science and Technology of China

Zhimin Huang
Adelphi University

ABSTRACT

The h index is widely used for academic evaluation as it incorporates the quantity and quality of scientific output in a simple way. However, after a lot of studies, scholars have found some drawbacks of the h-index, and proposed many corresponding indicators to improve h index. Although there are many h type indicators, few studies have taken into account the quality of citations. Most indices assume citations are equally important, but this assumption is obviously unpractical. A paper cited by top journals is more valuable than

Contemporary Perspectives in Data Mining, Volume 4, pp. 103–118
Copyright © 2021 by Information Age Publishing
All rights of reproduction in any form reserved.

cited by marginal journals. This chapter improves the h index by assigning weights to all citations based on the Eigenfactor scores of citation source, and proposes the h_{we} index (h type index weighted by Eigenfactor of citations).

This chapter validates the effectiveness of the h_{we} index by comparing the ranks of 20 journals in the field of Operations Research and Management Science. Five of these 20 journals are included in UTD24 journals listed by the University of Texas at Dallas. We analyze the differences and correlations among the h_{we} index, journal impact factors (JIF), Eigenfactor, and three h-type indices using Spearman correlation analysis and factor analysis. We find the h_{we} index is significantly correlated with Eigenfactor and h-type indices, but not with JIF, and the ranks of four UTD24 periodicals have been improved by h_{we} index compared with JIF. These results show that the index is an effective indicator that can reflect the real impact of journals, and h_{we} index makes an incremental contribution against the JIF.

We further propose $h_{we}(T)$ index to increase the discrimination of the index and to rank journals with same h_{we}.

INTRODUCTION

With the advent of the era of knowledge economy, science and technology are playing an increasingly important role in society. Academic journals, as the main carriers of research results, support most of the scientific information required by scientists. Scholars can publish their research achievement in journals. Research achievement could be studied and applied to practice, and scholars find problems and inspiration in practice. We can grasp research hotpots in various fields through scientific journals. Journal evaluation is an important research topic in the field of bibliometrics.

Academics generally use bibliometric indicators to evaluate the impact of journals, of which the most commonly used are journal impact factors (JIF). JIF is a quantitative indicator proposed by Garfield for journal assessment (Garfield, 1972). The meaning of the impact factor in any given year is the average number of citations per paper published in a journal during the two previous years (Garfield, 2006). Journal impact factor has become an international evaluation method, which is the basis for periodical division in Journal Citation Reports (JCR). With the widespread use of impact factor, some drawbacks have been summarized by scholars, including impact of self-citation, citations to non-citable items, English language bias, negligence of citation quality, and so forth (Falagas et al., 2008; Listed, 2006; Pinski & Narin, 1976; Postma, 2007; Seglen, 1997).

Pinski and Narin (1976) believed the biggest drawback of the impact factor is that all citations are considered equally important, without considering the quality of citation journals, and proposed a new method named popularity factor for the evaluation of journals weighted by citations. However, the popularity factor is rarely used because of its complexity.

Through continuous explorations by scholars, more new indicators have been applied to journal evaluation. Among new indicators proposed in recent years, Eigenfactor score (ES) has attracted popular attention. ES was designed to assess journals using a similar algorithm as Google's Page Rank, which takes into account not only the number of citations, but also the prestige of citation source (Bergstrom, 2007). The basic assumption of ES is that journals are more influential if they are often cited by other prestigious journals. In 2009, the ES was formally adopted by JCR as one of the basic indicators in the enhanced version of JCR. Journals' ES can be obtained directly from JCR.

Another indicator that draws wide attention is the h index. H index was proposed by Hirsch in 2005, and it is defined as "A scientist has index h if h of his or her Np papers have at least h citations each and the other (Np – h) papers have ≤ h citations each" (Hirsch, 2005, p. 16569). It means the first h publications received at least h citations each that ranked in decreasing order of the number of citations of each paper, and the h is the maximum number. These h papers form h-core (Rousseau, 2006).

H-index was quickly reported by *Nature* (Ball, 2005) and *Science* (Anon, 2005), and attracted wide attention in bibliometrics and information sciences since it was proposed. Bornmann and Daniel (2007) found 30 relevant articles published within one year after the present of h index. H index is a simple and original new indicator incorporating both quantity and quality of papers (Egghe, 2006a; Egghe & Rousseau, 2006). H index was originally designed to assess scientists' lifetime achievement, moreover, it can also be used to evaluate the impact of journals (Braun et al., 2006). The h index of a journal can be calculated as:

> Retrieving all source items of a given journal from a given year and sort-ing them by the number of times cited, it is easy to find the highest rank number which is still lower than the corresponding times cited value. This is exactly the h-index of the journal for the given year. (Braun et al., 2006, p. 170)

The journal's h index is calculated for a period impact, and impact of single year is the simplest case.

For the advantages and drawbacks of the h-index, scientists proposed some variants which can be also used to evaluate journals. Since the h-index does not highlight the value of highly cited papers, many scholars have proposed some h type indices. G-index proposed by Egghe (2006a, 2006b), is defined as the highest number g of publications that together received g^2 or more citations. G-index is the first h-type index. Wu's (2010) w index is defined as: If a researcher's w papers have received at least 10w citations each, and other papers have fewer than 10(w + 1) citations each, w is the researcher's w-index. G index and w index pay more attention to highly cited papers and more favorable to those scientists who have less output with highly cited. Some variants make full use of the information in the h core. A-index introduced by Jin (2006) is the average of citations the articles in the h-core received. A-index is sensitive to more widely cited papers in h-core. Jin et al. (2007) proposed R-index and AR-index. R-index is the square root of total number of citations of papers in the h-core, and it improves the sensitivity and discriminability of the papers in his h-core. AR-index is a modification of the R-index, which takes the age of the publications as a determinant. There are some other indices introduced to overcome some disadvantages of h-index, such as h (2) index (Kosmulski, 2006), b-index (Bornmann et al., 2007), h_{we} index (Egghe & Rousseau, 2008), e-index (zhang, 2009), and hbar index (Hirsch, 2010), and so forth.

Bornmann et al. (2011) studied the correlation between the h index and 37 h-type indices. In his review of the research literature and our literature study, we know there are more than 60 h type variants to date. Nevertheless, almost all these variants ignore the influence of citation source, and they evaluate journals under the assumption that all citations are equally important, just considering the number of citations each paper received. The influence of citation source should be a determinant in journals evaluation (Bergstrom, 2007; Pinski & Narin, 1976; SCImago Research Group, 2008). Yang et al. (2017) proposed an Extended H Index (EHI), which gives more weight to citations published in top journals. The EHI is the first h type index which considers the influence of citation source. It is more suitable for assessing outstanding scientists, but has low degree of discrimination between ordinary researchers, whose EHI are small and spread over a small range.

This chapter designs a new h type variant named h_{we} index (h type index weighted by Eigenfactor of citations), which takes into account influence of all citation sources, as a particularly scientific method for journal evaluation.

What Is the h_{we} Index

Journals are more influential if they are often cited by other prestigious journals (Bergstrom, 2007). This chapter weights citation by the value of the normalized Eigenfactor of citation source which can more effectively reflect the prestige of journals (Cantín et al., 2015; Sillet et al., 2012) and can reduce the disparity between citations in different disciplines.

The procedures for calculating the weights of citations are as follows. First, search for cited information of the evaluated journals from *Web of*

Science (WoS). Second, count the number of citations per paper received, and obtain the normalized Eigenfactor scores of each citation source from JCR. Third, calculate the average of normalized ES of all citation sources. Lastly, the weight of a citation is the value of ES of the citation source divided by the average of ES of all citation sources. We give an example to help understand the steps mentioned above. The average ES of all citation sources is A. One paper received n citations, and its ES vector is $E = (e_1, e_2..., e_n)$, e_1 represents the ES of the ith citation. Then the weight vector of citations of this chapter is $W = E / A = (\frac{e_1}{A}, \frac{e_2}{A},...,\frac{e_n}{A})$, $\frac{e_i}{A}$ represents the weight of the ith citation. This chapter $(e_1+e_2+...+e_n)/A$ has weighted citations.

The definition of the h_{we} index is as follows: if h_{we} of a journal's papers within a certain period of time has at least h_{we} weighted citations each and the other papers have fewer than $(h_{we} + 1)$ weighted citations, the journal's h_{we} index is h_{we}. It means if all papers published in a journal in that time period are ranked in descending order of the weighted number of citations they received, the first h_{we} papers receive at least h_{we} weighted citations each, and the h_{we} is the maximum possible value. These papers form core.

The h_{we} index maintains the unique advantage of the h index, it incorporates the quantity and quality of productivity, and also considers the importance of citations which is one of the determinants of journal evaluation. The h_{we} index is theoretically a more reasonable indicator to assess journals, but whether it is really effective still needs to be further tested. In the following studies, we choose three most interesting indices, impact factor, h index and Eigenfactor score, and two h type indices, g index and A index, to compare with h_{we} index.

Empirical Analysis

Data Set

The original data were collected in December 2017 from WoS and JCR. The journal's h index evaluates journals in a certain period of time, so we select the papers of each journal from 2014 to 2015 as a data set to calculate journals h index in 2016. In addition to the same data as h index, h_{we} index also needs EN of journals in which each citation published. JIF and ES of every journal in 2016 can be obtained directly from JCR.

This chapter selects 20 journals with high impact factors in the field of Operations Research and Management Science, and five of these journals are included in the 24 periodicals published by the University of Texas at Dallas (UTD24), which are top journals in the area of management.

Comparison of the Index With Other Methods

Table 7.1 shows the values of the indices of the 20 journals, and Table 7.2 shows the ranking of these journals based on the results of Table 7.1. We use abbreviation of journal to represent these journals. The journal titles in these two tables are abbreviations of the 20 journals. We mainly analyze Table 7.2.

Table 7.2 shows that the differences between the four h type indices are not very large, and we compare them in the following correlation analysis. We first analyze the ranking differences measured by impact factor, Eigenfactor score, h_{we} index and h index.

Table 7.1.
Values Of The 20 Journals

Journal title	JIF	ES	h_{we}	h	g	A
J OPER MANAG	5.207	0.0074	6	12	15	16.583
OMEGA-INT J MANAGE S	4.029	0.00877	13	18	24	27.667
EXPERT SYST APPL	3.928	0.0548	21	24	31	33.542
IEEE SYST J	3.882	0.00659	8	11	16	19.182
TRANSPORT RES B-METH	3.769	0.0121	9	15	20	21.867
INT J PROD ECON	3.493	0.02107	11	18	24	27.056
EUR J OPER RES	3.297	0.04474	19	21	34	41.381
TRANSPORT SCI	3.275	0.00499	6	7	15	24.875
TECHNOVATION	3.265	0.0039	4	8	10	10.625
DECIS SUPPORT SYST	3.222	0.01151	7	11	15	17.455
RELIAB ENG SYST SAFE	3.153	0.01283	9	14	20	23.286
TRANSPORT RES E-LOG	2.974	0.00799	9	12	15	15.833
MANAGE SCI	2.822	0.03578	11	14	18	20.643
J MANUF SYST	2.770	0.00287	5	9	11	12.556
NETW SPAT ECON	2.662	0.00189	3	7	10	12.429
COMPUT OPER RES	2.600	0.01675	14	15	18	19.000
SYST CONTROL LETT	2.550	0.01391	10	12	17	19.417
PROD OPER MANAG	1.850	0.0074	7	11	15	18.636
OPER RES	1.779	0.0125	7	9	12	14.222
M&SOM-MANUF SERV OP	1.683	0.00432	4	6	8	10.000

From the Table 7.2, we can see some differences among JIF, ES, h_{we} index and h index, but the rankings assessed by ES, h_{we} index and h index are partially the same. There are 4 journals with the same ranking measured by h_{we} index and h index, namely EXPERT SYST APPL, EUR J OPER RES, DECIS SUPPORT SYST and PROD OPER MANAG. Four journals are ranked in the same position by h_{we} index and ES, and they are EXPERT SYST APPL, EUR J OPER RES, TECHNOVATION and NETW SPAT ECON respectively.

Table 7.2 shows that the ranks measured by the JIF are very different from other methods. J OPER MANAG is measured as the most influential journal by the JIF, but the ES, h_{we} index, and h index rank it 13th, 15th, and 9th respectively. The ES, h_{we} index, and h index assess IEEE SYST J at the 15th, 11th and 12th places, but 4th place via JIF is very high. TRANSPORT SCI and TECHNOVATION are similar to the above two journals. They rank 8th and 9th according to the JIF, however, the ES, h_{we} index, and h index put them in the 15th to 18th positions, which are bottom of the rankings. The four journals mentioned above are ranked higher by the JIF, but ranked lower by the ES, h_{we} index, and h index. This suggests that the average impact of these journals are relatively high, but their influence in prestigious journals are still at a medium level. Contrary to the above, the rankings of some journals are lower through JIF than that assessed by the ES, h_{we} index, and h index. EXPERT SYST APPL is assessed as the most influential journal through the ES, index, and h index, indicating that this journal is highly prestigious and published many influential papers, but the JIF puts it in the 3rd position. Similarly, the rank of EUR J OPER RES via the ES, h_{we} index, and h index is 2nd, but by the JIF, EUR J OPER RES is ranked just 7th, which is clearly too low. Similar situation also exists in MANAGE SCI, COMPUT OPER RES, and SYST CONTROL LETT.

We focus on the five journals in the field of Operations Research and Management Science that included in UTD24, and they are J OPER MANAG, MANAGE SCI, PROD OPER MANAG, OPER RES, and M&SOM-MANUF SERV OP. The JIF, ES, h_{we} index, and h index put J OPER MANAG in the 1st, 13th, 15th, and 9th respectively, implying bias in the ES, index, and h index. The reason for the results may be due to the fact that fewer papers published in this journal between 2014 and 2015, and 82 papers have been published. If ranked according to the number of published papers, J OPER MANAG is ranked 19th among the 20 journals. MANAGE SCI is a recognized top academic journal, and is put 13th place by JIF. We also compare the rankings by the JIF of MANAGE SCI in the past five years, and find that the highest ranking was the 5th position in 2013, which are obviously too low. The rankings of MANAGE SCI by the ES, h_{we} index, and h index, which are 3rd, 5th, and 7th, have improved a lot compared with that of 13th via JIF. The ES, h_{we} index can better reflect the true prestige of MANAGE SCI. PROD OPER MANAG is ranked 18th among the 20 journals and 24th in Operations Research and Management Science assessed by JIF, and the ES, h_{we} index, and h index put it in the 13th, 13th, and 12th positions respectively. The rankings of OPER RES measured by the ES, h_{we} index, and h index raise 11, 7, 4 relative to 19th by the JIF. As for M&SOM-MANUF SERV OP, it is ranked last evaluated by the JIF, and the rank through the h index has not change, but rises to 17th and 18th assessed by the ES and index. The h_{we} index shows a strong consistency with the ES when evaluating these five top journals, because they all combine the influence of citations.

Table 7.2.
Ranks of the 20 Journals

Journal title	JIF	ES	h_{we}	h	g	A
J OPER MANAG	1	13	15	9	11	14
OMEGA-INT J MANAGE S	2	11	4	3	3	3
EXPERT SYST APPL	3	1	1	1	2	2
IEEE SYST J	4	15	11	12	10	10
TRANSPORT RES B-METH	5	9	8	5	5	7
INT J PROD ECON	6	4	5	3	3	4
EUR J OPER RES	7	2	2	2	1	1
TRANSPORT SCI	8	16	15	18	11	5
TECHNOVATION	9	18	18	17	18	19
DECIS SUPPORT SYST	10	10	12	12	11	13
RELIAB ENG SYST SAFE	11	7	8	7	5	6
TRANSPORT RES E-LOG	12	12	8	9	11	15
MANAGE SCI	13	3	5	7	7	8
J MANUF SYST	14	19	17	15	17	17
NETW SPAT ECON	15	20	20	18	18	18
COMPUT OPER RES	16	5	3	5	7	11
SYST CONTROL LETT	17	6	7	9	9	9
PROD OPER MANAG	18	13	12	12	11	12
OPER RES	19	8	12	15	16	16
M&SOM-MANUF SERV OP	20	17	18	20	20	20

Through the above analysis, we observe that the evaluation results made by each method are very different, and the difference between the JIF and other indicators is the largest. The h_{we} index demonstrates a degree of consistency with the ES and the h index. The ES and the h_{we} index can effectively reflect the prestige of journals.

To visualize the relations among the six measures, we compute a rank Spearman correlation analysis. The results are presented in Table 7.3.

From the Table 7.3, we know that the ES, h_{we} index, h index, g index, and A index are significantly related to each other at the 0.01 level. The highest correlation coefficient is found between the h index and the g index (0.944). The h_{we} index shows a high correlation with the h index (0.922), because they all take into consideration the number of articles and citations. As for the ES, it shows the highest correlation with the h_{we} index (0.898), because both indicators use the influence of citations as a determinant of the journal evaluation. The correlations between the h index, g index, A index and the JIF are significant at the 0.05 level. The ES and the h_{we} index are not significantly related to the JIF.

Table 7.3.
Spearman's Correlation Coefficients

	JIF	ES	h_{we} index	H index	G index	A index
JIF	1.000					
ES	0.152	1.000				
index	0.321	0.898**	1.000			
H index	0.516*	0.830**	0.926**	1.000		
G index	0.543*	0.823**	0.922**	0.944**	1.000	
A index	0.564**	0.708**	0.805**	0.790**	0.935**	1.000

* Significant at 5%.

** Significant at 1%

The higher correlations between the indicators indicate that they measure similar aspects of scientific impact, and nonsignificantly correlated indices can play an incremental role with each other (Bornmann et al., 2009; Bornmann et al., 2011). Through the correlation analysis, we know that the h_{we} index not only inherits the advantages of the h index, but also makes an incremental contribution against the JIF.

Bornmann et al. (2008) used an Exploratory Factor Analysis to discover factors that indicate how the h index and other eight h type indices calculated for the B.I.F. Applicants cluster. This chapter tries to find basic dimensions of the six indicators that indicate how they evaluate the 20 journals by using Exploratory Factor Analysis.

Table 7.4 shows that KMO=0.738, and the significance value of Bartlett's Sphericity test is 0.000, which less than the significance level, indicating that these data are suitable for factor analysis.

Table 7.4.
KMO and Bartlett's Test

Kaiser-Meyer-Olkin Measure of Sampling Adequacy		.738
Bartlett's test of Sphericity	Approx. Chi-Square	166.561
	df	15
	Sig.	.000

Table 7.5.
Total Variance Explained

	Initial Eigenvalues			Extraction Sums of Squared Loadings			Rotation Sums of Squared Loadings		
Component	Total	% of variance	Cumulative %	Total	% of variance	Cumulative %	Total	% of variance	Cumulative %
1	4.672	77.859	77.859	4.672	77.859	77.859	3.978	66.295	66.295
2	.949	15.814	93.673	.949	15.814	93.673	1.643	27.378	93.673
3	.230	3.841	97.514						
4	.101	1.683	99.198						
5	.040	.667	99.865						
6	.008	.135	100.000						

Table 7.5 shows that the cumulative variance of the first two factors is 93.673%, implying that the two factors can explain more than 93% of the total variance of the six indicators. The first two factors are extracted as main factors.

The results of analysis are presented in Table 7.6. It can be seen from the Table 7.6, that most communalities of the six indicators are greater than 90%, and only one is 85.7%, indicating that the factor analysis works well. The rotated factor loading matrix for the two main factors and the six indicators reveals that the ES, h_{we} index, h index, g index, and A index have a lager load on factor 1, where the loads of the ES and index reach

Table 7.6.
Rotated Component Matrix and Communalities

Index	Factor 1	Factor 2	Communality
JIF	.137	**.975**	.970
ES	**.967**	.014	.934
h$_{we}$ index	**.965**	.177	.963
h index	**.874**	.392	.918
g index	**.863**	.484	.979
A index	**.764**	.522	.857

Extraction Method: Principle Component Analysis.

Rotation Method: Varimax with Kaiser Normalization.

a. Rotation converged in 3 iterations.

more than 96%, and the JIF has a larger load on factor 2. From the Table 7.5, we know that factor 1 explains 66.29% and factor 2 explains 27.37% of the variance. Through the above analysis, we refer to factor 1 as "prestige of journal," and factor 2 as "popularity of journal." The h$_{we}$ index can reflect the academic value of journals well.

Further Improvement

It is impossible to further distinguish the ranking of journals with same index. To increase the degree of discrimination of the h$_{we}$ index, we further propose the h$_{we}$ (T) index, where the factor T represents the total number of weighted citations in the h$_{we}$ core. When the h$_{we}$ indices of journals are equal, the greater the value of T, the greater the influence of h$_{we}$ core, and the prestige of the journal will be higher.

This chapter illustrates the h$_{we}$ (T) index by distinguishing the journals with the same ranking in the Table 7.2. The results are presented in Table 7.7.

We take INT J PROD ECON and MANAGE SCI as an example. The h$_{we}$ indices of these two journals both equal 11, but the total number of weighted citations in core of MANAGE SCI (599.416) is greater than that of INT J PROD ECON (288.000). It is reasonable to believe that the prestige of MANAGE SCI is relatively high. From the Table 7.7, we can see that the h$_{we}$ (T) index can further distinguish journals with the same indices and improve the discrimination of the h$_{we}$ index.

Table 7.7.
The h$_{we}$ and h$_{we}$ (T) Indices Of Journals With Same

Journals	Values		Ranks	
	h$_{we}$	h$_{we}$ (T)	h$_{we}$	h$_{we}$ (T)
INT J PROD ECON	11	11(288.000)	5	6
MANAGE SCI	11	11(599.416)	5	5
TRANSPORT RES B-METH	9	9(549.272)	8	8
RELIAB ENG SYST SAFE	9	9(448.360)	8	10
TRANSPORT RES E-LOG	9	9(519.134)	8	9
DECIS SUPPORT SYST	7	7(249.863)	12	12
PROD OPER MANAG	7	7(61.849)	12	13
OPER RES	7	7(60.028)	12	14
J OPER MANAG	6	6(96.328)	15	15
TRANSPORT SCI	6	6(96.067)	15	16
TECHNOVATION	4	4(21.088)	18	18
M&SOM-MANUF SERV OP	4	4(20.174)	18	19

CONCLUSIONS AND DISCUSSION

Journal evaluation has always been a topic continuously explored by the academic community. Objective and effective evaluation methods can not only facilitate the development of academic journal, but also promote the progress of science and technology. Peer review is the most direct and effective evaluation method, but it requires a high level of knowledge of experts, and is difficult to implement on a large scale. Therefore, the bibliometrics method is generally used to evaluate journals, and one of the most commonly used measure is the citation analysis method. Journal impact factor, h index and Eigenfactor are the research hotpots in citation analysis, and they are also the most commonly used evaluation methods. The journal impact factor and the h index are based on the assumption that all citations are equally important. The Eigenfactor takes into account the value of different citations and is considered to be the most relevant to peer review and more effective than the impact factor. This indicator that combines the impact of citations is clearly more reasonable.

The h index has been a focus of research in the field of scientometrics since it was proposed, because it incorporates quantity and quality of scientific output. Many scientists designed variants of h index, and as far as we know, there are more than 60 h-type indices. These indices make up for the drawbacks of the h index in some ways, but seldom of them take the value of citations into consideration. This chapter improves the h index by assigning weights to all citations based on the Eigenfactor scores of citation source, and proposes the h_{we} index.

This chapter validates the effectiveness of the index by comparing the ranks of 20 journals in the field of Operations Research and Management Science. Five of these 20 journals are included in UTD24 journals. We analyze the differences and correlations among the h_{we} index, JIF, Eigenfactor, h index, g index, and A index using Spearman correlation analysis and factor analysis. We find the following important results:

1. The h_{we} index is significantly correlated with other h-type indices.
2. The h_{we} index is significantly correlated with the Eigenfactor, and the correlation between the h_{we} index and the Eigenfactor is most significant compared to other indices.
3. The h_{we} index loads the highest on the main factor 1 with coefficient of 0.965 among h-type indices.
4. Only the h_{we} index and the Eigenfactor are not related to the impact factor.
5. The ranks of the four UTD24 periodicals have been improved by h_{we} index compared with the impact factor.

These results shows that the h_{we} index is an effective indicator that can reflect the real impact of journals, and it is a useful complement to the bibliometric toolbox.

We further proposed h_{we} (T) index to increase the discrimination of the h_{we} index and to rank journals with same h_{we}. The factor T is the total citation number of the h_{we} core.

The h_{we} index proposed in this chapter is used for the evaluation of journals, and it is also suitable for scientists evaluation and other objects that h index can evaluate.

This chapter takes into account the quality of citations when proposing a new h-type index, and this provides a new thinking for the study of h index. In order to improve the discrimination of index, we first use the total number of citations of core output of journals which can be ranked when they have same index.

The conclusions of this chapter are mainly based on the data of specific field. Whether the conclusions have universal applicability remain to be further studied.

For future research, we have two directions:

1.   Collect more data to verify whether the conclusions of this chapter are universally applicable.
2.   Calculate the number of weighted citations for each paper within different citation time window to obtain the journal's h_{we} index. Study whether different citation time window will bring changes to the results.

Academic evaluation is a very complicated process. It is difficult to assess the true academic level of the evaluation objects using only a single indicator. In practical application, we can combine multiple indicators for comprehensive evaluation to make the assessment more accurate.

ACKNOWLEDGMENTS

The authors would like to thank National Natural Science Foundation of China (Grant nos. 71631006) and China National Social Science Foundation (Grant No.13CTQ042) for their financial support.

REFERENCES

Anon. (2005). Data point. *Science, 309*(5738), 1181.

Ball, P. (2005). *Index aims for fair ranking of scientists. Nature, 436,* 900. https://doi.org/10.1038/436900a

Bergstrom, C. (2007). Eigenfactor: Measuring the value and prestige of scholarly journals. *College & Research Libraries News, 68*(5), 314–316.

Bornmann L., & Daniel, H. D. (2007). What do we know about the h index? *Journal of the Association for Information Science and Technology, 58*(9), 1381–1385.

Bornmann L., Marx, W., & Schier, H. (2009). Hirsch Type index values for organic chemistry journals: A comparison of new metrics with the journal impact factor. *European Journal of Organic Chemistry, 2009*(10), 1471–1476.

Bornmann, L., Mutz, R., & Daniel, H, D. (2007). The b index as a measure of scientific excellence. *International Journal of Scientometrics, Informetrics and Bibliometrics, 11*(1.).

Bornmann, L., Mutz, R., & Daniel, H. D. (2008). Are there better indices for evaluation purposes than the h index? A comparison of nine different variants of the h index using data from biomedicine. *Journal of the American Society for Information Science and Technology, 59*(5), 830–837.

Bornmann, L., Mutz, R., Hug, S. E., & Daniel, H.-D. (2011). A multilevel meta-analysis of studies reporting correlations between the h index and 37 different h index variants. *Journal of Informetrics, 5*(3), 346–359.

Braun, T., Glänzel, W., & Schubert, A. A. (2006). Hirsch-type index for journals. *Scientometrics, 69*(1), 169–173.

Cantín, M., Muñoz, M., & Roa, I. (2015). Comparison between impact factor, eigenfactor score, and SCImago journal rank indicator in anatomy and morphology journals. *International Journal of Morphology, 33*(3).

Egghe, L. (2006a). An improvement of the h-index: The g-index. *ISSI Newsletter, 2*(1), 8–9.

Egghe L. (2006b). How to improve the h-index. *The Scientist, 20*(3), 15–16.

Egghe, L., & Rousseau, R. (2006). An informetric model for the Hirsch-index. *Scientometrics, 69*(1), 121–129.

Egghe, L., & Rousseau, R. (2008). An h-index weighted by citation impact. *Information Processing & Management, 44*(2), 770–780.

Falagas, M. E., Kouranos, V. D., Arencibiajorge, R., & Karageorgopoulos, D. E. (2008). Comparison of SCImago journal rank indicator with journal impact factor. *Faseb Journal, 22*(8), 2623–2628.

Garfield, E. (1972). Citation analysis as a tool in journal evaluation. *Science, 178*(4060), 471–479.

Garfield, E. (2006). The history and meaning of the journal impact factor. *Jama, 295*(1), 90–93.

Hirsch, J. (2010). An index to quantify an individual's scientific research output that takes into account the effect of multiple coauthorship. *Scientometrics, 85*(3), 741–754.

Hirsch, J. E. (2005). An index to quantify an individual's scientific research output. *Proceedings of the National academy of Sciences of the United States of America, 102*(46), 16569.

Jin, B. (2006). H-index: an evaluation indicator proposed by scientist. *Science Focus, 1*(1), 8–9.

Jin, B., Liang, L,, Rousseau, R., & Egghe, L. (2007). The R-and AR-indices: Complementing the h-index. *Chinese Science Bulletin, 52*(6), 855–863.

Kosmulski, M. (2006) A new Hirsch-type index saves time and works equally well as the original h-index. *ISSI Newsletter, 2*(3), 4–6.

Listed, N. (2006). The impact factor game. It is time to find a better way to assess the scientific literature. *Plos Medicine, 3*(6), e291.

Pinski, G., & Narin, F. (1976). Citation influence for journal aggregates of scientific publications: Theory, with application to the literature of physics. *Information Processing & Management, 12*(5), 297–312.

Postma, E. (2007). Inflated impact factors? The true impact of evolutionary papers in non-evolutionary journals. *Plos One, 2*(10), e999.

Rousseau, R. (2006). *New developments related to the Hirsch index. Science Focus, 1.*

SCImago Research Group. (2008). *Description of Scimago journal rank indicator.* https://www.scimagojr.com/SCImagoJournalRank.pdf

Seglen, P. O. (1997). Why the impact factor of journals should not be used for evaluating research. *BMJ, 314*(7079), 498–502.

Sillet, A., Katsahian, S., Rangé, H., Czernichow, S., & Bouchard, P. (2012). The Eigenfactor™ Score in highly specific medical fields: The dental model. *Journal of Dental Research, 91*(4), 329–333.

Wu, Q. (2010). The w–index: A measure to assess scientific impact by focusing on widely cited papers. *Journal of the American Society for Information Science and Technology, 61*(3), 609–614.

Yang, F., Zu, X., & Huang, Z. (2017). An extended h-index: a new method to evaluate scientists' impact. *Contemporary Perspectives in Data Mining, 3*, 135.

CHAPTER 8

A METHOD TO DETERMINE THE SIZE OF THE RESAMPLED DATA IN IMBALANCED CLASSIFICATION

**Matthew Bonas, Son Nguyen, Alan Olinsky,
John Quinn, and Phyllis Schumacher**
Bryant University

ABSTRACT

Classical positive-negative classification models often fail to detect positive observations in data that have a significantly low positive rate. This is a common problem in many domains, such as finance (fraud detection and bankruptcy detection), business (product categorization), and healthcare (rare disease diagnosis). A popular solution is to balance the data by random undersampling (RUS), that is, randomly remove a number of negative observations or random oversampling (ROS), that is, randomly reuse a number of positive observations. In this study, we discuss a generalization of RUS and ROS where the dataset becomes balanced, so that number of positive observations matches the number of negative observations. We also propose

Contemporary Perspectives in Data Mining, Volume 4, pp. 119–141

a data-driven method to determine the size of the resampled data that most improves classification models.

A METHOD TO DETERMINE THE SIZE OF THE RESAMPLED DATA IN IMBALANCED CLASSIFICATION

The idea and concept of imbalanced classification is very prominent in machine learning problems throughout the world today. Numerous studies have been done to further explore and understand the problems and difficulties associated with imbalanced data. Many real-world datasets deal with some degree of imbalance in the target variable of choice. Some examples would include rare-disease classification, bankruptcy prediction, and fraud detection. Traditional machine learning algorithms have difficulty in correctly identifying positive observations in these datasets. This chapter proposes a way to correct this problem of imbalanced data while simultaneously providing the reader the ability to determine the optimal size of the dataset to be used during machine learning analysis.

LITERATURE REVIEW

Imbalanced Datasets

A dataset is considered "imbalanced" when the variable of interest (i.e., the target variable) contains a significantly larger number of observations of one category compared to the number of observations for the other category(s). Chawla (2009) describes these types of datasets as gaining popularity since "real-world" problems are being solved using machine learning techniques and many of these types of problems have imbalanced classification categories. He gives a generic overview of data mining and the different concepts within the field of study. Concepts such as Receiver Operator Characteristic (ROC) curves, confusion matrices, precision and recall, and so forth, are discussed at length providing a thorough understanding of the content. These ideas are all used in determining how well one's model performs on the data. This is categorized under the overarching category of "performance measures." In general, these measures will most likely not yield accurate results on the imbalanced dataset when it comes to arranging new or test data in their respective classification categories. This leads Chawla to further discuss how to correct the imbalance of the dataset. Numerous resampling methods are discussed briefly throughout the chapter. Some sampling strategies include synthetic minority oversampling technique (SMOTE), Tomek Links, random oversampling

(ROS), random undersampling (RUS), and so forth. Each of these methods fall into the broader categories of undersampling, oversampling, or some combination of both.

This research highlights the problems and difficulties of working with imbalanced data. It is especially important due to the fact that a large portion of "real-world" data is imbalanced. Krawczyk (2016) provides many types of examples which are most likely to be considered imbalanced. Some of these examples include cancer malignancy grading, behavior analysis, text mining, software defect prediction, and so forth. Krawcyzk then describes the concept of binary imbalanced classification which is the most studied branch of learning in imbalanced data. It is defined as having a target variable with only two classes where the distinction between the two classes is well defined and where one class composes the majority of the dataset (Krawcyzk, 2016). It is further described how some of the more recent studies on imbalanced data have had class imbalances ratios of 1:4 or even 1:100. It further explains how there have not been as many studies done on problems with higher ratios of imbalance (i.e. 1:1000, 1:5000) (Krawcyzk, 2016). This higher imbalance is associated with problems such as fraud detection which is one of the problem of focus for this chapter.

Resampling Methods

As discussed above, there are many different ways one can rebalance an imbalanced dataset in order to create a more accurate model to classify new or test data. As discussed by Chawla et al. (2002) one of the techniques to resample the data is called synthetic minority oversampling technique (SMOTE). This technique is done by oversampling the minority class by creating new, "synthetic," examples until the minority class is equal to the majority class. The minority class is adjusted by taking each sample in the class and generating synthetic examples along the line segments that join the k minority class nearest neighbors (Chawla et al., 2002). The number of neighbors from the k nearest neighbors are randomly chosen according to the authors. SMOTE essentially forces the decision region of the minority class to be more generalized. Based on the generation of synthetic minority class observations within this method, SMOTE falls under the broader category of oversampling.

Another method of resampling applied to imbalanced datasets is called Tomek Link (T-Link). As defined by Elhassan et al. (2016), T-Link is an undersampling method developed by Tomek (1976) which is considered an enhancement of the nearest-neighbor rule (NNR). The algorithm is described as follows: "Let x be an instance of class A and y an instance of class B. Let d(x, y) be the distance between x and y. (x, y) is a T-Link, if for any instance z, d(x, y) < d(x, z) or d(x, y) < d(y, z). If any two examples are T-Link then one of these examples is noise or otherwise both examples are located on the boundary of the classes" (Elhassan et al., 2016). Essentially, the algorithm recognizes observations of different classes which are closely related and either: (1) Removes both observations creating a distinct boundary between the two classes or (2) removes only the majority class observations in these links to lower the overall imbalance ratio of the dataset. As stated above, Tomek Link is considered to be within the category of undersampling.

Finally, two more methods used to resample imbalanced datasets are called random undersampling (RUS) and random oversampling (ROS). As discussed by Shelke et al. (2017), these resampling methods are two of the simpler methods. Effectively, random undersampling is done by randomly selecting and removing majority class observations until the number of observations in the classification categories are equal. Random oversampling is accomplished by randomly selecting currently existing minority class observations and replicating them until the classification categories are equal (Shelke et al., 2017). Even though these methods are not as intensive as the aforementioned algorithms, they still manage to perform well in terms of rebalancing the dataset and even sometimes perform better in terms of balanced accuracy than the more complicated methods (Shelke et al., 2017).

For this chapter, each of the above resampling methods were used in an attempt to balance the datasets. Due to the sheer size of the dataset some of these methods were deemed impractical for they took an extensive amount of time to run and complete. After some initial tests, random undersampling was seen to be the highest performing model. All of the methods used, as well as their results, are explicitly explained and delineated below.

RESAMPLING METHODS

There are two very general categories when it comes to resampling an imbalanced dataset. These categories are undersampling and oversampling. There is an extensive number of methods under the umbrella of each of these categories which all have their different benefits and drawbacks. The method used in any given model is strictly at the discretion of the user. Each of these categories, as well as a method belonging to each, are defined below.

Undersampling

Undersampling is the broad category in which the model resamples the majority or larger class until it has the same number of observations as the minority class in the dataset. This resampling can be accomplished by randomly selecting and removing majority observations or by removing majority observations by some nearest neighbors/clustering algorithms. Random Undersampling (RUS) and Tomek-Links (T-Links) are among the more common undersampling techniques. For this analysis Tomek-Links was deemed impractical due to the amount of time the method took to complete. Only random undersampling was used to balance the dataset.

Random Undersampling. RUS is one of the simpler methods under the larger umbrella that is undersampling. Essentially, random undersampling is accomplished by randomly selecting majority class observations and removing them from the dataset. This then will create a smaller, balanced subset of the original dataset which has the benefit of a faster run-time relative to the original data, oversampling techniques, and even some other undersampling techniques. One major drawback of this technique is that by removing majority observations entirely from the dataset some valuable information could be lost which would then negatively affect the predictive power of the model overall (Shelke et al., 2017). Figure 8.1 is graphical representation of RUS.

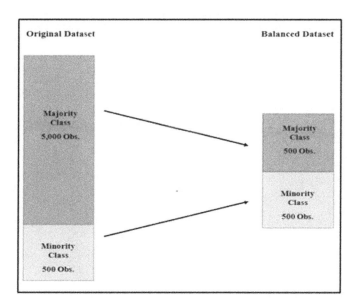

Figure 8.1.   Random undersampling.

Oversampling

Oversampling is the general category in which the model resamples the minority or smaller class of the dataset until it is balanced with the majority class. This resampling method can be accomplished either by random selection or synthetic creation. Some of the more common techniques in oversampling include synthetic minority oversampling technique (SMOTE) and random oversampling (ROS). For this analysis, SMOTE was deemed impractical due to the time it takes to run a model using this technique. Ultimately, random oversampling was the only oversampling technique chosen to balance the dataset.

Random Oversampling. Similar to random undersampling, random oversampling is one of the simpler models under its respective resampling category. It is virtually the exact opposite of RUS where it is accomplished by randomly selecting minority class observations and duplicating them until there are enough observations to make the dataset balanced. This will then create a larger dataset than the original dataset where now the original dataset is a subset of this resampled set. The major benefit of ROS is that none of the original information in the dataset is lost and that there is minimal bias injected into the dataset because existing observations are only being duplicated versus new observations being synthetically created. The downside of ROS is that it takes a longer time to run the model due

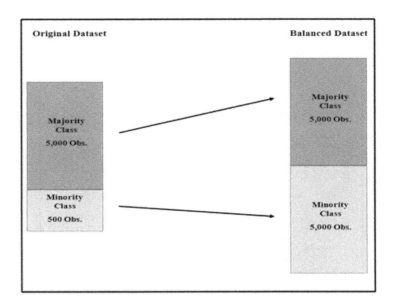

Figure 8.2.   Random oversampling.

to the larger size of the resampled dataset (Shelke et al., 2017). Figure 8.2 is a graphical representation of ROS.

Multiple Resampling

Multiple resampling is a concept that is essentially self-explanatory. It is completed by resampling the dataset an arbitrary number of times and creating a model using each of the resampled sets. Then, from these models, a majority vote is taken and the best prediction is used. This concept was used for this analysis and the arbitrary number of samples created was ten. We did not test to see whether the samples created perform better, or worse, as the number of samples increases and decreases. The value of ten samples was chosen strictly for the speed in which the samples were able to be created and used for modelling. A graphical representation of multiple resampling (using RUS) of a bankruptcy prediction dataset is depicted below in Figure 8.3.

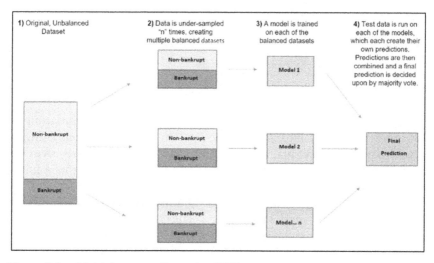

Figure 8.3. Multiple resampling using RUS.

PREDICTIVE MODELS

Three different supervised learning algorithms were initially used for predictions on the datasets. These algorithms are random forests, support vectors machines (SVMs), and decision trees. After initial testing with each model, only random forests were used with the resampling methods due to the performance as well as the faster run-time of these forests compared to the SVMs and decision trees. Each of the models are described in detail below.

Decision Trees

Decision trees are a useful alternative to regression analysis used in predictive modeling. The trees are created through algorithms that find various ways of splitting up a dataset into segments based on nominal, ordinal, and continuous variables (Christie et al., 2009). These segmentations are useful in that the effects of variables are developed hierarchically when the model is being developed. It enables researchers to look beyond the simple x and y relationship since we are able to incorporate more variables. We can now describe data outcomes in the context of multiple influences. Decision trees fall short in that they can lead to spurious relationships in the data. There is also the chance of overfitting the data, that is, over classifying the data so that the learning algorithm continues to develop hypotheses that reduce training set error at the cost of an increased test set error (Christie et al., 2009).

Random Forests

The fundamental idea behind a random forest is a combination of decision trees joined into a single model. The purpose of combining is to show that an individual decision tree may not be the most accurate predictive method but a combination of models would bring us a more accurate prediction (Biau, 2012). The forest is created by generating an arbitrary number of trees and then taking a majority vote over the forest to determine the best model for the data. The creation of random forests is advantageous as it fixes the decision tree issue of overfitting. Additionally, random forests have less variance than a single decision tree in general (Biau, 2012). As stated previously, random forests consistently performed better than any other machine learning algorithm chosen.

Support Vector Machines

Support vector machines were introduced by Vladimir N. Vapnik and Alexey Ya Chervonenkis in 1963. An SVM builds a model for the data that separates it into two categories by producing a clear gap between the categories. New data is then mapped onto that same space and predicted which category it belongs to by determining which side of the gap it falls (Cortes & Vapnik, 1995). Later on in 1992, Bernhard E. Boser, Isabelle M. Guyon, and Vapnik developed a way to create nonlinear classifiers by applying the kernel trick to maximum-margin hyperplanes (Boser et al., 1992). The kernel trick transforms the data and then based on these transformations the SVM finds the new optimal boundary. However, it should be noted that working in a higher-dimensional feature space increases the generalization error of SVMs, although given enough samples, the algorithm still performs well (Jin & Wang, 2012).

PERFORMANCE MEASURES

For all the following subsections refer to Figure 8.4.

Sensitivity

$$Sensitivity = \frac{TP}{(TP+FN)}$$

Sensitivity, also known as the true positive rate, is a measure of the model's ability to correctly determine positive observations. A sensitivity score can range from zero (0) to one (1). In context, a score of zero indicates the model failed to identify any positive observations. Conversely, a sensitivity score of one indicates the model correctly identified all the positive observations in the data.

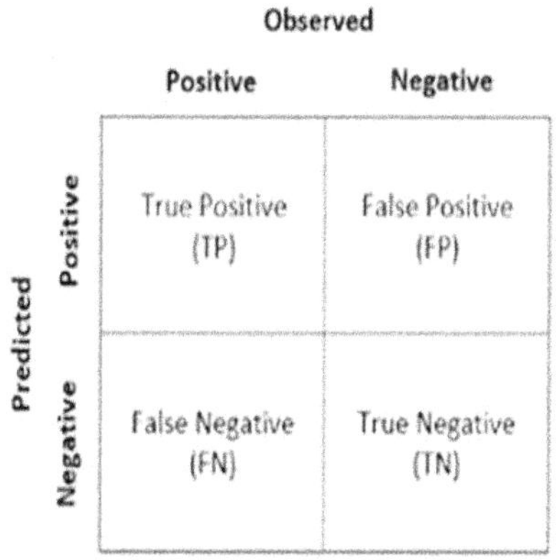

Figure 8.4.   Confusion matrix.

Specificity

$$Specificity = \frac{TN}{(TN+FP)}$$

Specificity, also known as the true negative rate, is a measure of the model's ability to correctly determine negative observations. Like sensitivity, a specificity score can range anywhere from zero to one. A score of zero indicates that the model failed to classify any negative observations. Alternatively, a sensitivity score of one shows that the model correctly identified all the negative observations in the data.

Balanced Accuracy

$$Balanced\ Accuracy = \frac{\left[\left(\frac{TP}{(TP+FN)}\right)+\left(\frac{TN}{(TN+FP)}\right)\right]}{2} = \frac{sensitivity+specificity}{2}$$

Balanced accuracy is another measure that can be used to determine how well a model performs. Balanced accuracy measures the model's ability to correctly predict and identify both positive and negative observations in the data. Because of this, balanced accuracy is a more powerful and telling measure when analyzing imbalanced datasets.

The benefits of using balanced accuracy can be seen by conceptualizing the following case. If a model was being trained on a dataset of 100 potential cancer patients where 99 were classified as "not cancer" and 1 was classified as "cancer" then the model would have problems classifying the "cancer" patients due to the high imbalance of the data. If this model predicted all observations as "not cancer" then the model would have an overall accuracy of 99%. However, when you consider the sensitivity, specificity, and balanced accuracy scores the models performance would be abysmal. The model would have a sensitivity of zero because it failed to identify the only "cancer" observation in the dataset. Additionally, even though the specificity score would be 1 because all negative observations were identified correctly, the balanced accuracy score would be 0.5 or 50%. Having a balanced accuracy of 50% when predicting cancer patients clearly indicates the deficiency in a model which deals with potentially life threatening issues and also shows how it is a better measure of the model's predictive power than that of the standard accuracy measures.

DATASETS OVERVIEW

Three datasets were chosen to be used in this experiment. Each dataset has a different size and different ratio of imbalance between the majority and minority classes. Additionally, each dataset contains different types of data. The datasets were chosen for the diversity in their data and their structures, and are described in detail below.

Credit Card Fraud

The credit card fraud was taken from the Kaggle database. The dataset consists of credit card transactions made over the course of two days in September 2013 by European cardholders. The target variable is "Class" denoted as 1 for fraudulent transaction or 0 for non-fraudulent transactions. The other variables in the dataset are "Amount" defined as the amount spent during the transaction, and "Time" defined as the time elapsed since the first transaction in the dataset. There are 28 other interval variables which remain anonymous due to confidentiality reasons. It was noted that the variables were chosen using principal component analysis (PCA). In a dataset of 284,807 transactions, approximately 0.173% of them were fraudulent. Figure A1 in Appendix A depicts the percentage of observations in each of the target categories.

Car Accidents

This dataset includes car accidents in the United States from 2015. The target variable "DEFORMED" shows whether the car had major damage (denoted as 1) or minor/no damage (denoted as 0). The dataset has approximately 81,000 observations with 28 variables. Some of the other variables in the dataset include the month, day, time, weather, age of the driver, gender of the driver, object hit/collided with, state, etc. Figure A2 in Appendix A shows the percentage of observations belonging to each target class in this dataset.

Basketball Hall of Fame

This dataset was retrieved from basketball references and it includes a historical list of players from the National Basketball Association (NBA) and whether each player was inducted into the Hall of Fame (HOF). The target variable "Hall_of_Fame" has a value of one if the player is in the HOF and a value of zero if the player is not in the HOF. The dataset has approximately 1450 observations with 38 variables including the target variable. Some of the other variables in the dataset are team, games played, field goals made, field goals attempted, steals, blocks, turnovers, total points, all-star, most valuable player, and so forth. Figure A3 in Appendix A shows the percentage of observations belonging to each of the target categories in the dataset.

METHOD

In this test, the concept of multiple resampling was ultimately expanded upon. Instead of using only RUS or ROS for the multiple resampling, the datasets were balanced at intermediate ranges between the values of ROS and RUS (and even some values below RUS). For this, a parameter for the size of the dataset was coined as the "S-Value" for size-value. Using Python version 3.7.2, code was created where multiple resampling would be completed over a specified range of these S-Values. For context, when the S-Value equals the number of observations in the majority class then this would be a case of ROS, for the minority class is now equivalent (in terms of number of observations) to the majority class. Similarly, when the S-Value is equal to the number of observations in the minority class then this would be a case of RUS. The question being answered using this method of balancing is: *What S-value produces the highest balanced accuracy for a specific dataset?*

These new, balanced datasets were then used in conjunction with a random forest with 15 trees. As stated previously, random forests were the only supervised learning model used for analysis, for initial testing showed they produced higher balanced accuracies and ran quickly in comparison to other predictive models. The function created for this analysis has 11 parameters. The parameters for the "multresamp" function are listed and defined in Table 8.1.

This function will only produce a single balanced accuracy value for the specified "r" parameter. To gain a better understanding as to how the balanced accuracy changes, we ultimately decided to create a new function ("graphmult") which uses the first function and loops over a range of "r" parameter values and produces a graph of the balanced accuracy versus the S-value of the dataset. The graph(s) produced gave us a better understanding into how the balanced accuracy may fluctuate across different S-values. The parameters for the "graphmult" function are listed and defined in Table 8.2. The parameters which are shared with the "multresamp" function are only defined in Table 8.1. Both functions are depicted in Appendix B.

RESULTS

With the functions above, we were able to analyze and determine the optimal S-value which produced the highest balanced accuracy for each of the datasets. The results for each respective dataset are explained in depth below. All the figures referenced in the following subsections can be found in Appendix C.

Table 8.1.
Multiple Resampling Function Parameter (Multresamp)

Parameter	Definition
X_train0	Training dataset including all negative observations without the target variable column
y_train0	Training dataset of only the target variable column including only negative observations
X_train1	Training dataset including all positive observations without the target variable column
y_train1	Training dataset of only the target variable column including only positive observations
X_test	Testing dataset without the target variable column
y_test	Testing dataset of only the target variable column
n0	Number of observations in X_train0 and y_train0
n1	Number of observations in X_train1 and y_train1
ntest	Number of observations in X_test and y_test
r	Value to be multiplied by n0 to determine the S-Value of the dataset (i.e. $r = 1$ is ROS)
loop	Number of time multiple resampling is performed (default value of 10)

Table 8.2.
Graphing Function Parameters (Graphmult)

Parameter	Definition
startloop	Specifies the first value in the loop (default value of 1)
endloop	Specifies the last value in the loop
increment	Determines the number of loops to be completed. For example, if startloop = 1 and endloop = 10 then increment = 10 would loop over the "r" values 0.1, 0.2, ..., 0.9, 1.0

Credit Card/Car Accident Datasets

When used with the function both the credit card dataset and the car accident dataset had similar results. With the credit card dataset, it can be seen from Figure 8.A4 and Figure 8.A5 that as the S-value of the data approached random oversampling (the point furthest right in Figure 8.A4) that the balanced accuracy decreased or remained stagnant. Conversely, as the S-value approached random undersampling (depicted as the blacked dashed line in Figure 8.A9) the Balanced accuracy appears to increase. Both previous observations remain true as well when analyzing Figure 8.A6 which depicts the results for the car accident dataset. This clear trend of higher balanced accuracy towards the RUS point led us to question whether there was any sort of overfitting occurring with the data, especially at the larger end of the spectrum for S-values.

Overfitting

Overfitting is defined as when the model learns the noise and extensive detail of the training data to the point that it might negatively impact the performance of the model overall. Because the model learns the randomness and noise of the training data, any new data being tested will ultimately suffer, for this noise and randomness does not apply to the testing data. To test if our models were overfitting we used the original training data in place of the testing data. Essentially what this means is that we are testing a model using the data it was trained with. For overfitting to be present we should see a balanced accuracy score near or equivalent to 1.0 or 100%.

Credit Card Overfitting. For the credit card dataset, it can be seen in Figure 8.A7 that as the size of the dataset increases, or the S-value increases, that the balanced accuracy tends to approach 1.0. This trend towards a perfect balanced accuracy score occurs rather early or at the lower end of the S-value spectrum. Figure 8.A8 shows how the balanced accuracy remains nearly perfect at the larger S-values of the credit card dataset. Both the figures highlight that as the S-value approaches the ROS point that the model begins to (rapidly) overfit the data. This then translates as to why, in Figure 8.A4, the balanced accuracy of the testing portion of the credit card dataset was so low, relative to the rest of the graph, close to the ROS point.

Car Accident Overfitting. Like the credit card dataset, it can be seen in Figure 8.A9 that as the S-value of the dataset increases the balanced accuracy appears to approach a value of 1.0. Again, this chart explicitly shows that as the S-value approaches the ROS point that the model begins to overfit the data. However, unlike the credit card dataset, this trend towards a perfect balanced accuracy score occurs at a slow pace. This overfitting is why, in Figure 8.A6, the testing portion of the dataset performs worse at the larger S-values.

Basketball Hall of Fame Dataset

The basketball dataset had surprisingly different results than both the credit card and the car accident datasets. When the resampling function was used on the basketball dataset it appeared that there was no significant difference between the balanced accuracy scores at the ROS and RUS points or any other S-value points in between. These results are shown in Figure 8.4. While the balanced accuracy does fluctuate, this fluctuation only covers approximately 1.0%. One could attribute these fluctuations to the natural noise and randomness of the data. We then tested to see whether the model was overfitting this dataset as well. We found that, according to Figure 8.5, overfitting did not occur for this dataset. Figure 8.5 is like Figure 8.4 in that the balanced accuracy only fluctuates a minor amount across all the S-values chosen. Since the graph does not approach 1.0 anywhere we could not conclude that overfitting occurred.

This analysis with the basketball dataset led us to hypothesize that random oversampling may only be practical for smaller datasets, and even then may not necessarily perform better than random undersampling. From these results it appears that as the size of the dataset gets larger, the chance of overfitting with random oversampling also increases.

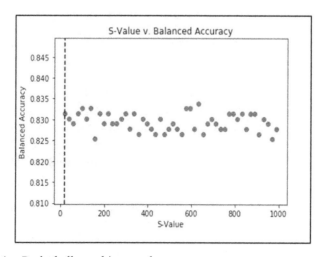

Figure 8.4.   Basketball graphing results.

CONCLUSION/FUTURE STUDY

This analysis, using a modified form of multiple resampling, gave us many insights into the concept of resampling imbalanced datasets. From our

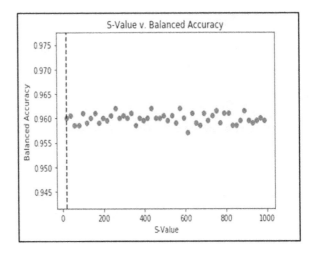

Figure 8.5. Basketball overfitting results.

results, we concluded that when the datasets were resampled, that the majority of them had their highest balanced accuracy scores around the area of random undersampling. We also found that as the S-value of the resampled data increases, that the balanced accuracy of the model tends to decrease in most cases. Both of these observations led us to the believe that random undersampling performed as well as, if not better than, random oversampling in all cases. From our results it could be argued that one should almost always choose to use RUS over ROS because of how well it performs, in terms of balanced accuracy. Additionally, one could argue to choose RUS because of how quickly it completes its runs, relative to the speed of ROS. Overall, we were able to create and use a function in Python that will allow any user to take any dataset and find the optimal S-value for which the balanced accuracy is at its maximum. Again, the code for both the multiple resampling function as well as the graphing function can be found in Appendix B.

The problem and concept of resampling imbalanced datasets will always be ongoing. To expand on what we showed above one could attempt to use different resampling methods such as SMOTE or T-Links during the multiple resampling procedure. This would allow for an interesting comparison as to how this might affect what the optimal S-value is for a given dataset. Additionally, it may be interesting to explore how these different resampling methods may compare to ROS in terms of overfitting the model at larger, or smaller, S-values. Lastly, one could explore how different supervised learning algorithms compare to the current random forest chosen in the analysis. It would be interesting to see how SVM or a neural network alters the optimal S-value of the dataset.

REFERENCES

Biau, G. (2012, April). Analysis of a random forests model. *Journal of Machine Learning Research, 13*, 1063–1095.

Boser, B. E., Guyon, I. M., & Vapnik, V. N. (1992, July). A training algorithm for optimal margin classifiers. In *Proceedings of the fifth annual workshop on computational learning theory* (pp. 144–152). ACM.

Chawla, N. V. (2009). Data mining for imbalanced datasets: An overview. In *Data mining and knowledge discovery handbook* (pp. 875–886). Springer.

Chawla, N. V., Bowyer, K. W., Hall, L. O., & Kegelmeyer, W. P. (2002). SMOTE: synthetic minority over-sampling technique. *Journal of Artificial Intelligence Research, 16*, 321–357.

Christie, P., Georges, J., Thompson, J., & Wells, C. (2009). *Applied analytics using SAS Enterprise Miner*. SAS Institute.

Cortes, C., & Vapnik, V. (1995). Support-vector networks. *Machine Learning, 20*(3), 273–297.

Elhassan, A., Aljourf, M., Al-Mohanna, F., & Shoukri. M. (2016). Classification of imbalance data using Tomek Link (T-Link) vombined with random under-sampling (RUS) as a data reduction method. *Global Journal of Technology Optimization*. https://doi.org/10.4172/2229-8711.S1111

Jin, C., & Wang, L. (2012). Dimensionality dependent PAC-Bayes margin bound. *Advances In Neural Information Processing Systems, 2*, 1034–1042.

Krawczyk, B. (2016). Learning from imbalanced data: Open challenges and future directions. *Progress in Artificial Intelligence, 5*(4), 221–232.

Tomek, I. (1976). An experiment with the edited nearest-neighbor rule. *IEEE Transactions on systems, Man, and Cybernetics, 6*(6), 448–452.

Shelke, M. M. S., Deshmukh, P. R., & Shandilya, V. K. (2017, April). A Review on imbalanced data handling using undersampling and oversampling technique. *International Journal of Recent Trends in Engineering & Research, 3*(4). https://doi.org/10.23883/IJRTER.2017.3168.0UWXM

APPENDIX A

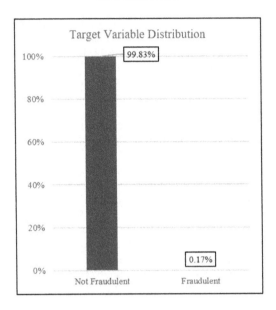

Figure A1.   Credit card target variable distribution.

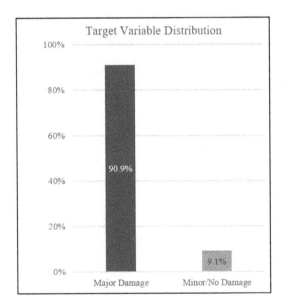

Figure A2.   Car Accident target variable distribution.

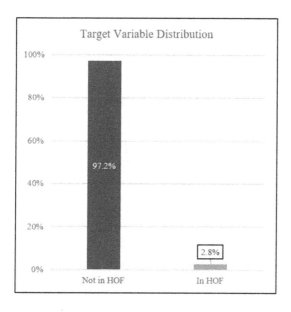

Figure A3.   Basketball target variable distribution.

APPENDIX B

Multiple Resampling Function

```
def multresamp(X_train0, y_train0, X_train1, y_train1, X_test, y_test, n0, n1, ntest, r, loop = 10) :

    mc = np.empty((ntest, loop-1))
    rf = RandomForestClassifier(n_estimators = 15)
    alpha = r * n0

    for i in range(1, loop)  :
        X_train00 = X_train0[np.random.randint(n0, size = round(alpha))]
        y_train00 = y_train0[np.random.randint(n0, size = round(alpha))]
        X_train01 = X_train1[np.random.randint(n1, size = round(alpha))]
        y_train01 = y_train1[np.random.randint(n1, size = round(alpha))]
        X_trainunder = np.concatenate((X_train00, X_train01),0)
        y_trainunder = np.concatenate((y_train00, y_train01),0)

        rf.fit(X_trainunder, y_trainunder)
        mc[:,i-1] = rf.predict(X_test)

    cc = np.round(np.mean(mc, axis = 1), decimals = 0)

    return(balanced_accuracy_score(y_test, cc))
```

Graphing Function

```
def graphmult(startloop = 1, endloop, increment, X_train0, y_train0, X_train1, y_train1, X_test, y_test, n0, n1, ntest, loop = 10) :
    rvb = np.empty((endloop,2))
    fixloop = endloop+1


    for i in range(startloop ,fixloop) :
        rvalue = i/increment
        rvb[i-1,0] = rvalue
        rvb[i-1,1] = multresamp(X_train0, y_train0, X_train1, y_train1, X_test, y_test, n0, n1, ntest, rvalue, loop)

    ba = rvb[:,1]
    rvals = rvb[:,0]

    plt.scatter(rvals*n0, ba)
    plt.xlabel('S-Value')
    plt.ylabel('Balanced Accuracy')
    plt.title('S-Value v. Balanced Accuracy')
    plt.axvline(x=n1, color='black', ls = 'dashed')
    plt.show()
```

APPENDIX C

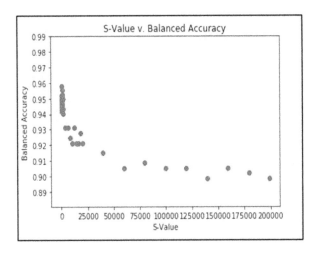

Figure A4.   Credit Card graphing results—1.

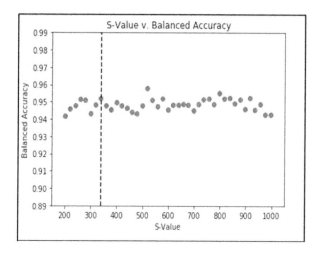

Figure A5.   Credit card graphing results—2.

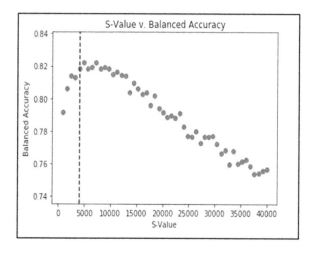

Figure A6.   Car accidents graphing results.

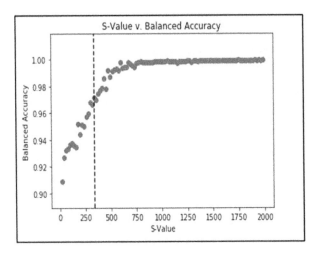

Figure A7.   Credit card overfitting results—1.

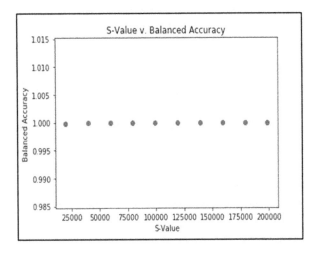

Figure A8.   Credit card overfitting results—2.

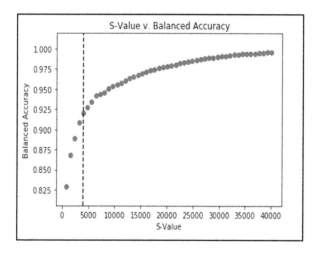

Figure A9.   Car accident overfitting results.

CHAPTER 9

PERFORMANCE MEASURES ANALYSIS OF THE AMERICAN WATER WORKS COMPANY BY STATISTICAL CLUSTERING

Kenneth D. Lawrence and Stephen K. Kudbya
New Jersey Institute of Technology

Sheila M. Lawrence
Rutgers, The State University of New Jersey

PEER GROUPS AND EXECUTIVE COMPENSATION

In order to attain a more clear understanding of executive compensation, a standardized method supported by statistical techniques can provide a sound base. An essential item to initiate the process is to identify a peer group or a collection of other corporations to measure themselves against when calculating the executive compensation package.

An example of a peer group in the water works industry is a list of comparative companies selected by the Compensation Committee of America Water Works, institutional investors of America Water Works, and advisors

Contemporary Perspectives in Data Mining, Volume 4, pp. 143–150

to American Water Works. They sever as a market benchmark for evaluating executive pay levels and company pay design (Dikolli et al., 2010; Hermalin & Weisback, 1998).

Benchmarking against a peer group helps the Compensation Committee determine the competitiveness of its compensation plan, as well as to demonstrate the alignment the of compensation and performance relative to peers. When selecting a peer group, consider industry size and other qualitative factors such as competition for talent, global footprint or unique situational factors that may drive the selection of particular financial variables (Anderson, 2003; Brochet, et al., 2014; Stewart & Roth, 2001).

In order to establish an objective sample in selecting peer groups for particular companies, the rules of the Security and Exchange Commission requires a Compensation Committee to provide the names of the corporations it uses in its peer groups and to provide an extensive description of its executive compensation philosophy.

From a corporate leadership perspective, evaluating performance as the individual or corporate level can be expressed:

1. exceeds expectation
2. meets expectations
3. does not meet expectations.

These performance standards become more meaningful when they are evaluated relative to a peer group; thus, minimize "rater bias" and allowing better allocation of compensation dollars. For executives, employers, their peer groups and relative performance are anchored to external benchmarks.

PERFORMANCE MEASURES DATA ELEMENTS

Corporate Valuation Measures

The first set of variables for the peer measurement involve the valuation of corporations. Finding the value of a corporation is no simple task. It is an essential component of effective corporate management. The nine measures are as follows:

1. Market Capitalization
2. Enterprise Value
3. Trailing P/E
4. Forward P/E
5. PEG Ratio

6. Price/Sales
7. Price/Book
8. Enterprise Value/Revenue
9. Enterprise Value/EBITA

Profitability

The next set of variables involve the corporation's profit. The two measures that are used are:

1. Profit Margin
2. Operating Margin

MANAGEMENT EFFECTIVENESS

The success or failure of a corporation depends on how effective its operations and strategies are. The corporation's management is responsible for propelling the future growth in the right direction including such elements as investment and utilization of technology, hiring and developing its workforce, and so forth. It is also responsible for administering and controlling the corporation's activities and accounting for its results. Ineffective management can result in the failure of the corporation.

The following are the measures of managerial effectiveness used in this peer analysis:

1. Return of Assets
2. Return on Equity

Income Statement

The following are the income statement items used in this peer analysis:

1. Revenue
2. Revenue per Share
3. Quarterly Revenue Growth
4. Gross Profit
5. EBITA
6. Diluted EPA
7. Quarterly Earnings Growth

Balance Sheet Measures

The balance sheet measures that are used in this peer study:

1. Total Cash
2. Total Cash per Share
3. Total Debt
4. Total Debt/Equity
5. Current Ratio
6. Book Value per Share

Cash Flow Statement

1. Operating Cash Flow
2. Levered Free Cash Flow

SEGMENTATION OF AMERICAN WATER WORKS PEER COMPANIES BY CLUSTERING

A series of cluster runs were made on the data of the peer companies of American Water Works. Each observation corresponded to a particular peer company, and each of the 29 variables was one of the performance measures of each company.

The clustering needs were made for each of the five clustering methods over the seven agglomeration measures. Thus, 35 separate runs were made. Tables for each of the sets of runs will be set out as part of the discussion.

1. Alliant Energy
2. Atmos Energy
3. Center Point Energy
4. CMS Energy
5. Evergy
6. Ever Source Energy
7. MDU Resources
8. NI Source
9. OGE Energy
10. PPL
11. Pinnacles West Capital
12. Scanna

13. UGI Corporation
14. WEC Energy

AGGLOMERATIVE CLUSTERING AND DENDOGRAMS

A dendogram is a visual representation (e.g., the graphs) of the results of hierarchical procedure in which each object is arranged on one axis and the other axis portrays the steps in the hierarchical procedures. Starting with each objective represented as a separate cluster, the dendogram shows graphically how the clusters' are combined at each step of the procedure until all are contained in a single cluster.

Hierarchical clustering procedures involve a series of n-1 clustering decisions, where n equals the number of observations that combine the observations into a hierarchy of a tree-like structure. In the agglomerative methods of clustering each observation starts out as its own cluster. (Aggarawal & Roddy, 2014); Hair et al., 1998; Henning et al., 2016; Lattin et al, 2003)

The agglomeration methods follows a simple repetitive process. It begins with all observations as their own individual cluster, so that the number of clusters equals the number of observations. A similarity measure is used to combine the two most similar clusters into a new-cluster, thus reducing the number of clusters by one. This process is repeated again using the similarity measure to combine the two most similar clusters into a new cluster. This process continues at each step, combining the two most similar clusters into a new cluster. This process is repeated a total n–1 times until all are contained in a single cluster.

Table 9.1.
Euclidean Linkage Method

Distance Measure	Companies		
	Cluster 1	Cluster 2	Cluster 3
A – Centroid	1,2,3,4,5,7,8,9,10,11,12,14	6	13
B – Average	1,2,3,4,5,7,8,9,10,11,12,14	6	13
C – Complete	1,2,3,4,5,7,8,9,10,11,12,14	6	13
D – McGulty	1,2,3,4,5,7,8,9,10,11,12,14	6	13
E - Median	1,2,3,4,5,7,8,9,10,11,12,14	6	13
F – Single	1,2,3,4,5,7,8,9,10,11,12,14	6	13
G – Ward	1,2,3,4,5,7,8,9,10,11,12,14	6	13

Table 9.2.
Pearson Linkage Method

Distance Measure	Companies		
	Cluster 1	Cluster 2	Cluster 3
A – Centroid	1,2,3,4,5,6,7,8,9,10,12,14	11	13
B – Average	1,2,3,4,5,6,7,8,9,10,12,14	11	13
C – Complete	1,2,3,4,5,6,7,8,9,10,12,14	11	13
D – McGulty	1,2,3,4,5,6,7,8,9,10,12,14	11	13
E - Median	1,2,3,4,5,6,7,8,9,10,12,14	11	13
F – Single	1,2,3,4,5,6,7,8,9,10,12,14	11	13
G – Ward	1,2,3,4,5,6,7,8,9,10,12,14	11	13

Table 9.3.
Manhattan Linkage Method

Distance Measure	Companies		
	Cluster 1	Cluster 2	Cluster 3
A – Centroid	1,2,3,4,5,7,8,9,10,12,14	6	13
B – Average	1,2,3,4,5,7,8,9,10,12,14	6	13
C – Complete	1,2,3,4,5,7,8,9,10,12,14	6	13
D – McGulty	1,2,3,4,5,7,8,9,10,12,14	6	13
E - Median	1,2,3,4,5,7,8,9,10,12,14	6	13
F – Single	1,2,3,4,5,7,8,9,10,12,14	6	13
G – Ward	1,2,3,4,5,7,8,9,10,12,14	6	13

Table 9.4.
Squared Linkage Method

Distance Measure	Companies		
	Cluster 1	Cluster 2	Cluster 3
A – Centroid	1,2,3,4,5,7,8,9,10,11,12,14	6	13
B – Average	1,2,3,4,5,7,8,9,10,11,12,14	6	13
C – Complete	1,2,3,4,5,7,8,9,10,11,12,14	6	13
D – McGulty	1,2,3,4,5,7,8,9,10,11,12,14	6	13
E - Median	1,2,3,4,5,7,8,9,10,11,12,14	6	13
F – Single	1,2,3,4,5,7,8,9,10,11,12,14	6	13
G – Ward	1,2,3,4,5,7,8,9,10,11,12,14	6	13

Table 9.5.
Squared Pearson Linkage Method

	Companies		
Distance Measure	Cluster 1	Cluster 2	Cluster 3
A – Centroid	2,3,4,5,6,7,8,9,10,12,13	1,13	11
B – Average	2,3,4,5,6,7,8,9,10	1,13	11
C – Complete	2,3,4,5,6,7,8,9,10	1,13	11
D – McGulty	2,3,4,5,6,7,8,9,10	1,13	11
E - Median	2,3,4,5,6,7,8,9,10	1,13	11
F – Single	1,2,3,4,5,6,7,8,10,12,14	9,13	11
G – Ward	1,2,3,5,6,7,10,11,12	4,8,9,14	13

After the results of five sets of clustering, each of which had seven different distance measures, it is clear that the company #13 was significantly different from other companies in the American Water Works peer group (1,2,3,4,5,7,8,9,10,12,14) in all 35 cases. Also, in all 28 cases, company #6 was significantly different form all other American Water peer group (1,2,3,4,5,7,8,9,,10,11,12,14). Finally in 7 cases, company #11 in the American Water Works peer group was significantly different from the other companies (1,23,3,4,5,6,7,8,9,10,12,14).

Clustering analysis enabled a robust examination of peer organizations, considering numerous criteria. The results of this methodology provide a mechanism to optimize a peer group approach (e.g., refine the peer group to those more similar). With this information, other analytic methods can be used to address additional queries.

REFERENCES

Aggarwal, C., & Roddy, C. (2014). *Data clustering algorithms and applications.* CRC Press.

Anderson, R., & Reeb, D. (2003). An empirical investigation of the relationship between corporate ownership structures and their performance. *Journal of Finance, 58,* 1301–1323.

Brochet, F., Loumioto, M., & Serafin, E. (2014). Speaking of the short-term: disclosure horizons and capital market outcome. *Review of Accounting Studies, 20*(13), 1122–1163.

Dikolli, S., Meyer, W., & Narada, W. (2014). CEO tenure and the performance turnover relation. *Review of Accounting Studies, 19*(1), 281–327.

Hair, J., Anderson, R., Tatham, E., & Black, W. (1998). *Multivariate data analysis, 5e.* Prentice-Hall.

Henning, C., Meila, M., Murtogh, F., & Rocci, R. (2016). *Handbook of cluster analysis.* CRC Press.

Hermalin, B., & Weisbach, M. (1998). Endogenously chosen boards of directors and their monitoring the CEO. *American Economic Review, 88*(1), 96–118.

Lattin, J., Carrol, J., & Green, P. (2003). *Analyzing multivariate data.* Duxbury.

Stewart, W., & Roth, B. (2001). Risk propensity difference. *Journal of Applied Psychology, 86,* 145–152.

ABOUT THE AUTHORS

THE EDITORS

Kenneth D. Lawrence is a Professor of Management Science and Business Analytics in the School of Management at the New Jersey Institute of Technology. Professor Lawrence's research is in the areas of applied management science, data mining, forecasting, and multicriteria decision-making. His current research works include multicriteria mathematical programming models for productivity analysis, discriminant analysis, portfolio modeling, quantitative finance, and forecasting/data mining. He is a full member of the Graduate Doctoral Faculty of Management at Rutgers, The State University of New Jersey in the Department of Management Science and Information Systems and a Research Fellow in the Center for Supply Chain Management in the Rutgers Business School. His research work has been cited over 1,750 times in over 235 journals, including: Computers and Operations Research, International Journal of Forecasting, Journal of Marketing, Sloan Management Review, Management Science, and Technometrics. He has 375 publications in 28 journals including: *European Journal of Operational Research, Computers and Operations Research, Operational Research Quarterly, International Journal of Forecasting and Technometrics.* Professor Lawrence is Associated Editor of the *International Journal of Strategic Decision Making* (IGI Publishing). He is also Associated Editor of the *Review of Quantitative Finance and Accounting* (Springer-Verlag), as well as Associate Editor of the *Journal of Statistical Computation and Simulation* (Taylor and Francis). He is Editor of *Advances in Business and Management*

Forecasting (Emerald Press), Editor of *Applications of Management Science* (Emerald Press), and Editor of *Advances in Mathematical Programming and Financial Planning* (Emerald Press).

Ronald K. Klimberg, is a Professor in the Department of Decision Systems Sciences of the Haub School of Business at Saint Joseph's University. Dr. Klimberg has published 3 books, including his *Fundamentals of Predictive Analytics Using JMP,* edited 9 books, over 50 articles and made over 70 presentations at national and international conferences. His current major interest include multiple criteria decision making (MCDM), multiple objective linear programming (MOLP), data envelopment analysis (DEA), facility location, data visualization, data mining, risk analysis, workforce scheduling, and modeling in generation. He is currently a member of INFORMS, DSI, and MCDM. Ron was the 2007 recipients of the Tenglemann Award for his excellence in scholarship, teaching, and research.

THE AUTHORS

William Asterino is a supply chain financial analyst at a large healthcare company where his role consists primarily of using data mining techniques and Bayesian models to calculate and validate manufacturing costs. He is a proponent of programming and automation to enhance efficiency in the workplace. William graduated from Saint Joseph's University in 2019 with degrees in Business Intelligence & Analytics and Accounting. His research interests include health & fitness (optimizing a nutrition and exercise regimen to an individual), and assessment of player performance in sports via location tracking data. William aims to continue his education in the future through Georgia Tech's MS in Analytics.

Joel Asay, ABD, is an Assistant Professor of Business Analytics & Information Systems in the Williams College of Business at Xavier University. He has been a faculty member at Xavier since 2015. His primary teaching responsibilities are in the areas of database and R programming. He holds a BS in Economics from Brigham Young University, an MBA from Xavier University, and is in process to complete his DBA from Creighton University in 2020.

Matthew Bonas earned his BS in applied mathematics and statistics from Bryant University in 2019. He is currently a PhD student in the applied and computational mathematics and statistics department at the University of Notre Dame. Prior to attending Notre Dame, he was a Data Specialist for the Governor's Workforce Board of Rhode Island, a branch within the Department of Labor and Training. His current research is in the development of spatio-temporal statistical models for environmental applications.

Kathleen Campbell Garwood is an Assistant Professor in the Department of Decision & System Sciences and has been a member of the faculty at SJU since 2004. She advanced to Assistant Professor in 2014 after finishing her PhD at Temple University in Statistics. Her teaching interests include data mining and modeling with the goal of introducing real data information analysis techniques to students. Her research interests include community engaged research (addressing real world problems using analytics in the classroom), data visualization (the time and energy saved by making clear, understandable, and meaningful visuals for CPS), rank order comparisons (specifically collegiate and sustainability rankings), STEM research and the role that gender plays in both younger students as well as collegiate major selection in STEM fields, and modeling applications in real world settings (such as Fe y Alegría: Bolivia and Colonial Electric), on communication of the overall structure and state of the CPS in a form that can be interpreted quickly and precisely to help identify issues and concerns

An-Sing Chen is Distinguished Professor of Finance at National Chung Cheng University, Taiwan. He is currently the Chief Investment Officer for the university. He is a board director of MACAUTO INDUSTRIAL CO., LTD, a leading global supplier of automotive interior products. He was previously the Dean of the College of Business and the Chair of the Department of Finance at National Chung Cheng University. Professor Chen received his PhD in Business Economics from Indiana University. His current research interest is in the use of artificial intelligence in investing, portfolio management and trading of financial assets. He has published articles in numerous academic journals, including *Journal of Banking & Finance, Journal of Forecasting, International Journal of Forecasting, Computers and Operations Research, European Journal of Operational Research, Journal of Investing, Quantitative Finance*, among others. His recent research on oil price thresholds has been cited by the financial website MarketWatch.

Meng-Chen Hsieh received the PhD in Statistics from Stern School of Business, New York University in 2006. She is currently an Assistant Professor in the Department of Information Systems, Analytics, and Supply Chain Management at the Norm Brodsky College of Business of Rider University. Prior to her academic work she held the positions of Postdoctoral Researcher at the IBM T. J. Watson Research Center, Quantitative Researcher at Morgan Stanley and Credit Suisse, and Senior Statistician at Activision. Her research interests include developing financial econometrics models for transactional-data, inventory forecasting for supply chain management, and prescriptions of data-driven optimal decisions.

Cuihua Hu is a vice professor of information management and information system at School of Information Management, Shanghai Lixin University of Accounting and Finance in China. She holds a doctorate in management. Her research interest includes finance information management, business data analysis, information system audit and decision science. She has published more than 30 articles in many academic journals.

Zhimin Huang is Professor of Operations Management in the School of Business at Adelphi University. He received his PhD in Management Science from The University of Texas at Austin. His research interests are mainly in supply chain management, data envelopment analysis, distribution channels, game theory, chance constrained programming theory, and multicriteria decision making analysis.

Stephan Kudyba is an Associate Professor at NJIT's Tuchman School of Management. He has published numerous journal articles in data mining and analytics, as well as in such areas as supply chain management, finance, and health care.

Sheila M. Lawrence is an Assistant Teaching Professor in the School of Management and Labor Relations at Rutgers, The State University where she earned her PhD. She has 113 technical publications in statistics, supply chain, management science, and forecasting. Her more than 30 years of work experience includes The State of New Jersey, PSE&G, Hoffmann-LaRoche, and AT&T.

Mark T. Leung is Chair of the Department of Management Science and Statistics at the University of Texas at San Antonio. He received his PhD in operations management (with minor in decision sciences) from Indiana University. His research interests are in financial forecasting and predictive modeling, artificial intelligence, neural networks and machine learning, and supply chain analytics. He has taught a variety of management science and operations technology courses in undergraduate and graduate levels and received numerous awards including twice for the University of Texas System Chancellor Awards, the UT System Regents Teaching Award, and the Col. Jean Migliorino and Lt. Col. Philip Piccione Endowed Research Award. His research has been published in a spectrum of journals such as *Computers and Operations Research, Decision Sciences, Decision Support Systems European Journal of Operational Research, Expert Systems with Applications, International Journal of Production Economics, International Journal of Forecasting,* and *Journal of Banking and Finance.* He is on the editorial boards of several journals and constantly provides op-ed to media channels in recent years.

Aishwarya Mohanakrishnan graduated with MS in Information Systems from Penn State, Harrisburg. She has several years of software development and technical experience in India and has been working as a Data Science Analyst at SiriusXM Holdings–Pandora Media in the United States. Her area of interests are business analytics, data science and machine learning in the different industrial sectors and how data is used as the source of truth to validate and make decisions across different companies.

Son Nguyen earned his master's degree in applied mathematics and doctoral degree in mathematics, statistics emphasis, both at Ohio University. He is currently an assistant professor at the department of mathematics at Bryant University. His primary research interests lie in dimensionality reduction, imbalanced classification, and statistical learning. In addition to the theoretical aspects, he is also interested in applying statistics to other areas such as finance and healthcare.

Alan Olinsky is a professor of mathematics and computer information systems at Bryant University. He earned his PhD in Management Science from the University of Rhode Island and his research interests include multivariate statistics, management science, business analytics, and data mining. He is past president of the Rhode Island Chapter of the American Statistical Association and has appeared several times as an expert witness in statistical matters at hearings and trials.

Dinesh R. Pai is an Associate Professor of Supply Chain Management at the School of Business Administration at Penn State Harrisburg. He earned his PhD at Rutgers, The State University of New Jersey. His research interests are in the areas of supply chain management, business analytics, and performance evaluation. His research publications have appeared in *Health Systems, Information Technology and Management,* and *Expert Systems with Applications,* among others.

Shaotao Pan is a senior software engineer/data scientist at Visa Inc. He received his Master of Science degree in Statistics and Data Science from the University of Texas at San Antonio. His research interests are in statistical modeling, predictive analysis, deep learning, and big data. He is the coauthor of a number of book chapters and papers.

Jamie Pawlukiewicz is Professor and former Chair of the Department of Finance in the Williams College of Business at Xavier University. He has been on faculty at Xavier since 1989. His primary teaching responsibilities are in the areas of valuation and financial modeling while his research interests include the areas of valuation and hedge fund performance. He holds a BS in Business and Economics, an M.S. in Economics, and a PhD in Finance from the University of Kentucky

John Quinn is a Professor of Mathematics at Bryant University and has been teaching there since 1991. Prior to teaching, Professor Quinn was a mechanical engineer at the Naval Underwater Systems Center (now the Naval Undersea Warfare Center) in Newport, Rhode Island. He received his PhD degree from Harvard University in 1987 and 1991, respectively. Professor Quinn has had articles published in multiple areas. He has done previous research in mathematical programming methods and computable general equilibrium models in economics. He currently does research in simulation models and in data mining applications, including the prediction of rare events.

Phyllis Schumacher is a professor emeritas of mathematics at Bryant University. She taught both graduate and undergraduate students at Bryant for over 40 years. She has served as the Chair of the Math Department and the Coordinator of the Actuarial Statistics Program. Phyllis earned her PhD in Statistics from the University of Connecticut. Her research interests include data analysis, mathematics education, gender issues in mathematics and technology and the application of statistics to psychology. She has published articles on these topics in professional journals in psychology, mathematics and business education including *Computers in Human Behavior, The Journal of Education for Business,* and *Primus.*

Gregory Smith is Professor and Chair of the Business Analytics and Information Systems Department in the Williams College of Business at Xavier University. He holds a PhD in Business Information Technology with a focus in Management Science from the Virginia Polytechnic Institute and State University. A former analyst and benefits consultant, Dr. Smith has been on faculty at Xavier since 2006. At Xavier, he focuses his teaching and research in the areas of computer-based data mining applications, data modelling and business statistics, along with their associated ethical implications.

Girish H. Subramanian is Professor of Information Systems at the School of Business Administration at Penn State Harrisburg. His research interests are in software engineering, global software development and enterprise systems. His research publications have appeared in *Communications of the ACM, Decision Sciences, Journal of Management Information Systems, IEEE*

Transactions on Software Engineering, Journal of Computer Information Systems, and other journals.

Zu Xiya is a graduate student at School of Management, University of Science and Technology of China, majoring in Management Science and Engineering. Her research direction is information system and decision science. She has published a book chapter in *Contemporary Perspectives in Data Mining Volume 3*.

Feng Yang is a full professor of management at School of Management, University of Science and Technology of China. He holds a doctorate in management. His research interest includes supply chain management, data envelopment analysis, business data analysis, information system and decision science. He has published more than 90 articles in many international academic journals such as *EJOR, DSS, IJPR,* and others.

CPSIA information can be obtained
at www.ICGtesting.com
Printed in the USA
LVHW061242140121
675623LV00004B/23

9 781648 021435